Comhairle Cathrach
Bhaile Átha Cliath
Dublin City Council
Brainse Ràth Maonais
Rathmines Branch
Fón / Tel: 4973539
Due Date
Due Date
Due Date
D0508971

BY THE SAME AUTHOR

Beyond Babel: New Directions in Communications
The Half-Parent: Living with Other People's Children
Who's Afraid of Elizabeth Taylor?
Married and Gay: An Intimate Look at a Different Relationship
Maggie: The First Lady
The Pope and Contraception: The Diabolic Doctrine
D.H. Lawrence: The Story of a Marriage
Yeats's Ghosts: The Secret Life of W.B. Yeats
Nora: The Real Life of Molly Bloom
Rosalind Franklin: The Dark Lady of DNA
Freud's Wizard: The Enigma of Ernest Jones
George Eliot in Love

READING THE ROCKS

How Victorian Geologists Discovered the Secret of Life

Brenda Maddox

BLOOMSBURY PUBLISHING
LONDON • OXFORD • NEW YORK • NEW DELHI • SYDNEY

BLOOMSBURY PUBLISHING
Bloomsbury Publishing Plc
50 Bedford Square, London, WC1B 3DP, UK

BLOOMSBURY, BLOOMSBURY PUBLISHING and the Diana logo are
trademarks of Bloomsbury Publishing Plc

First published in Great Britain 2017
This edition published 2018

A catalogue record for this book is available from the British Library

ISBN: HB: 978-1-4088-7958-0; TPB: 978-1-4088-7960-3;
PB: 978-1-4088-7955-9; eBook: 978-1-4088-7961-0

2 4 6 8 10 9 7 5 3 1

Typeset by NewGen KnowledgeWorks (P) Ltd., Chennai, India
Printed and bound in Great Britain by CPI Group (UK) Ltd, Croydon CR0 4YY

To find out more about our authors and books visit
www.bloomsbury.com and sign up for our newsletters

For Laura

CONTENTS

FOREWORD

Why write about Victorian geologists? For me, the simple answer was 'George Eliot'. Having accepted an invitation to write a biography of the Victorian novelist,[1] I was intrigued to learn that Mary Ann Evans (her real name) had been an ardent geologist. She was introduced to the young science by her life partner, George Henry Lewes. Their enthusiasm for hammering the rocks on holidays at Tenby and the Isle of Wight brought alive to me the excitement of the mid-nineteenth century, when geology was new, as in many ways was investigative science itself. The scientific journal *Nature* began in 1869. Only slightly older was the *Economist*, the weekly public-affairs magazine founded in 1843, whose early contributors included the geologist and philosopher Herbert Spencer, who escorted Mary Ann Evans to theatres and concerts but would not marry her because he thought her ugly.

Drawn by Eliot and Lewes, I was fascinated by the period when many people – clergymen not least – pursued the new science of geology not only because they loved it but because it opened a window on the earth's ancient past. They showed great courage in facing the conflict between geology and Genesis that immediately presented itself. The rocks and fossils being dug up showed that the earth was immeasurably old, not the creation of six days as the Bible claimed. Moreover, the fossil evidence revealed upward progress in the changing forms of life. It was Charles Darwin's early career as a geologist that led him to recognise the direction of evolution. Observing the enormous variety of living things and their constant struggle for existence, Darwin drew the magisterial conclusion: 'whilst this planet has gone cycling on according to the fixed law of gravity, from so simple a beginning endless forms most beautiful and most wonderful have been, and are being, evolved'.[2] Darwin traced a clear path from fossils to man. His book published in 1859, *On the Origin of Species*, would prove to be the most influential document of the nineteenth century.

Researching the history of geology, I could see a line extending straight from 1807, when the Geological Society of London was formed, to 1830 and the publication of the popular classic *Principles of Geology* by Charles Lyell, to 1859 when Darwin's *Origin* appeared. (Darwin was a member of the Geological Society from 1836.) Luckily for a biographer, I found these early geologists to be superb letter-writers. Professor Martin Rudwick, the Cambridge geological historian, has observed that the speed and reliability of the early-nineteenth-century post gave 'scientific correspondence an immediacy and vitality that it had never had in earlier generations and, arguably, that it has never had since'.[3] He has pointed out that these scientists hoarded their letters as a medium of scientific exchange. This book is the richer for their hoarding.

I was also moved to write this book because geology changed my life. At least 'Geology 1' at Harvard did. Decades ago when I was an undergraduate there (in the women's college, then called Radcliffe) I found that Harvard in its wisdom would not allow anyone to work in just one of the major disciplines – humanities, social sciences, physical science – without taking at least one course in the other two. My field of concentration was English literature and, like many others, football players not least, I took what seemed the easy way out of the science requirement: geology. Not for nothing was 'Geology 1' known as 'Rocks for Jocks'. To my surprise, I enjoyed it, particularly the field trips to see the Spouting Rock at Marblehead and the tracks of 200-million-year-old three-toed dinosaurs in the Connecticut River Valley (this valley, believed to have been a subtropical swamp, is recognised as one of the world's richest grounds for discovering tracks of Late Triassic dinosaurs).

When after graduation I became a journalist on the well-regarded daily newspaper the *Quincy Patriot Ledger*, I found that Geology 1 had made me a science writer. No one else would touch the subject, while I found it easy. The estimable *Ledger*, science-conscious in the age of Sputnik, enlisted scientists from Harvard and the Massachusetts Institute of Technology to come to Quincy and give the staff lectures on the origin of life and the future of space research. I wrote these up for the paper which, out of its conviction that its readers deserved to be informed that science was news, made space on the front page.

Geology also found me a husband. In 1958 I took the summer off from the *Ledger* and went to Europe. In Paris I stayed with a young woman whose brother worked at the Commissariat à l'Energie Atomique and who was going to a conference in Geneva. When I mentioned the atomic energy conference in a letter to the *Ledger*, the managing editor John Herbert wrote back and said: 'Why not cover it for us? Quincy makes ships [the Bethlehem Steel Shipyard at Fore River was then the headquarters of the Bethlehem Shipbuilding Corporation]. We should know about atom-powered ships.'

Off I went to Geneva armed with a press card – entry credentials that sat me alongside distinguished journalists such as John Finney of the *New York Times* and Mary Goldring of the *Economist*, and allowed me to sit in at sessions at the Palais des Nations. I wrote daily articles for the *Ledger* – transmitted by telex – which I typed myself, explaining science's hopes for nuclear fusion: that some day, energy might be produced by fusing atoms (fusion) instead of splitting them apart (fission). The *Ledger* ran these stories under my byline 'Brenda Murphy, *Patriot Ledger* Staff Reporter', with headlines such as 'Fusion Goal Still Far Off', 'Russia Claims World's Largest A-Power Plant Put into Action', 'Nuclear Ships Worth Cost Today Only in Rare Cases', and (one of my lighter pieces) 'Scientists Are Avid Doodlers'.

At one morning's press conference, I felt someone looking at me across the crowded room. He was, I knew from the sharp questions he threw, John Maddox, science correspondent of the *Manchester Guardian*.

Two years and two months later we were married and living in London. He went on to become editor of the science journal *Nature*, while I worked briefly at Reuters, then for many years at the *Economist*, which, true to its non-specialist tradition, assigned me to write about such disparate subjects as telecommunications, local government, family law and Ireland. My trips to Ireland, north and south, reignited my college interest in James Joyce. In time I decided to perform a journalistic investigation of the facts I knew must be available and to write the biography of Joyce's unexamined wife, Nora Barnacle.

From then on I followed the biographer's trail, mainly in pursuit of literary figures. In time, after a scientific detour for *Rosalind Franklin: The Dark Lady of DNA*, it led me to George Eliot and her fascination with rocks. What follows is the story of those whose discoveries led them to a bigger truth than they had been looking for.

I

THE ABYSS OF TIME

An era marked by the youth of Britain's queen was the first to grasp the vast age of the earth. By 1837, when the eighteen-year-old Victoria ascended the throne, the new science of geology had shown the planet's hidden past extending back beyond human imagination, certainly beyond human existence.

In 1830 in his bestseller, *Principles of Geology*, Charles Lyell, a young Scottish barrister turned geologist, brought to a popular audience the information – indeed, the news – that the earth was *ancient*. The history of nature, Lyell said, showed evidence of 'elevation of the land above the sea' and 'seas and lakes filled up'. There were signs too that some of 'the lands whereon the forests grew have disappeared or changed their form'. He remarked on traces of plants belonging 'to species which for ages have passed away from the surface of our planet'.[1] In older rocks the causes of destruction had been obscured 'by the immense lapse of ages during which they have acted'. He was also one of the first to proclaim that rock strata were like an ancient manuscript that could be read as history.

Lyell's great book received immediate acclaim among the upper classes (who could afford to buy copies), thanks to its originality and clarity. Later editions reached the new reading public that had been created by the arrival of steam-powered printing, the electric telegraph and railways, as well as by the rise in popular literacy. The expanding rail network, with trains travelling at the magical speed of a mile a minute, carried newspapers, journals and magazines the length of the land to an eager audience waiting to snap them up.[2] Reader demand led to a wide network of circulating libraries, such as Mudie's, offering new books on loan for a yearly subscription fee, and also to railway booksellers such as W. H. Smith whose first bookstall opened at Euston Station in 1848.

While serialised novels by famous authors such as Charles Dickens and George Eliot were most sought after, books on science were also much in demand.[3] Lyell's three-volume *Principles* started the trend for science books aimed at the ordinary public. The book was so successful that its publisher, John Murray, repeatedly reissued it until the twelfth edition in 1875, the year of the author's death.

When *Principles of Geology* first appeared, the weekly *Spectator*, at that time known for its liberal stance, welcomed Lyell's judgement of the morally improving effect of geology. 'There are other investigations which more nearly affect our social happiness than the philosophy of geology,' it stated, 'but perhaps there is none which in an indirect manner produce a more wholesome and beneficial effect upon the mind . . . After the perusal of Mr. Lyell's volume, we confess to emotions of humility, to aspirations of the mind, to an elevation of thought, altogether foreign from the ordinary temper of worldly and busy men . . .' There are 'sermons in stones and tongues in brooks,' concluded the *Spectator*, 'but they want an interpreter: that interpreter is the enlightened geologist. Such a man is Mr. Lyell.'[4]

The British public were already in love with science. The first balloon crossing of the English Channel had been made in 1785, portable 'chemical chests' were on sale in Piccadilly, and for an expenditure of between six and twenty guineas affluent amateurs could buy the glassware and chemical ingredients for experiments at home, such as an air pump, electrical apparatus and a small heating unit. Open lectures on science regularly drew large crowds to London's Royal Institution on Albemarle Street. Founded in 1799, the RI, as it was (and is still) known, was dedicated to 'the application of science to the common purposes of life'. The institution's Friday night discourses drew such an audience that Albemarle Street was made London's first one-way street due to the crowds arriving in carriages.

The RI's most popular speaker was the boyishly handsome Cornish chemist Humphry Davy. In 1797 Davy had suddenly became fascinated by chemistry because, according to the cultural historian Richard Holmes, the subject was becoming 'the Romantic science par excellence. The last of the old alchemy was being replaced by true experiments, accurate measuring and weighing, and a new understanding of the fundamental processes of combustion, respiration and chemical bonding.'[5] Davy had recently arrived in London from the Pneumatic Institution in Bristol, where he had discovered the anaesthetic possibilities of nitrous oxide. He went on to study what he called 'galvanism' – electricity produced by chemical action.

In his inaugural lecture at the RI on 25 April 1801, Davy thrilled his audience with spectacular bursts of sparks and explosions (starting the institution's tradition of vivid displays that continues today). He offered a vision of what new scientific discoveries would mean to mankind, telling his enthralled audience: 'The composition of the atmosphere, and the properties of gases, have been ascertained; the phenomena of electricity have been developed; the lightnings have been taken from the clouds; and lastly, a new influence [nitrous oxide] has been discovered, which has enabled man to produce from combinations of dead matter effects which were formerly occasioned only by animal organs.'[6]

Moving on to the subject of geology, he gave ten lectures at the RI in 1805–06 on the subject, treating his audience (which included a large number of women even though, as was the custom of the time, they were not allowed to be members of the institution) to an explanation of the difference between the theories of 'Plutonism', as advocated by the Scot, James Hutton – who believed the extreme heat at the centre of the earth had pushed up and created the continents – and the 'Neptunism' of the German Abraham Werner – who argued that a primordial ocean had once blanketed the earth, and that 'the mountains, deserts, and farm lands had precipitated out of the receding water of the ocean . . . and the land on which humans lived was revealed.'[7]

In November 1806, with the naturalist Joseph Banks in the chair, Davy gave a lecture on the nature of electricity that created an international sensation. He did not strive for understatement. 'Till this discovery, our means were limited; the field of pneumatic research had been exhausted, and little remained for the experimentalist except minute and laborious processes,' he declared. 'There is now before us a boundless prospect of novelty in science; a country unexplored, but noble and fertile in aspect; a land of promise in philosophy.'[8]

Many drew inspiration from Davy's lectures, including Michael Faraday, a future star of the institution, who was employed as Davy's assistant. Faraday's research into the magnetism created by an electric current led to the invention of a dynamo that generated electricity, and was the foundation for electric motor technology. A new world opened. Faraday's Christmas lectures would become widely popular and start another RI tradition still alive today.

By 1815 Davy was best known for his invention of a lamp for miners that removed much of the great danger from the essential and dirty business of mining. Working in darkness at depths as great as 600 feet, men had to carry oil lamps or candles, risking the constant possibility of setting off an explosion in the surrounding flammable gases. The Davy Lamp, as it was known, was a formidable contribution to a vital, growing (yet still dangerous) industry. Davy became 'Sir Humphry', receiving his honour in 1812 as the first scientist to be knighted since Isaac Newton in 1705.

In *Consolations in Travel*, published posthumously in 1830, Davy ventured to examine some of the wider questions of science, faith and geology. Charles Lyell read the book and quoted from it in *Principles*. Though he was unsure of Davy's argument on the progressive development of organic life, the work helped convince him that man was of comparatively recent origin.

Lyell's own interest in geology had first been aroused by reading in his father's library Robert Bakewell's seminal work, published in 1813: *An Introduction to Geology (Illustrative of the General Structure of the Earth, Comprising the Elements of the Science, and an Outline of the Geology and Mineral Geography of England)*. However, it was during his time at Oxford, between 1816 and 1819, that Lyell learned the 'enlightened geology' that he would pour into his *Principles*.

Although reading classics at Exeter College, Lyell attended the lectures of the most famous clerical geologist of his time: the flamboyant Reverend William Buckland. Considering the challenge that geology threw down at religion, it may seem surprising in retrospect that so many early geologists were clergymen. Yet geology had been introduced at Oxford expressly in order to prepare the many students about to enter the Church to defend religion against science.

In his entertaining lectures, which were extracurricular (that is, not required for a degree), Buckland passed on to his student Lyell his passion for geology and fossils. (Fossils, from the Latin *fossilus*, had been defined since 1569 as the remains of animals or plants dug up from the earth.) It was Buckland who introduced Lyell to the debate on Plutonism and Neptunism, the two alternative theories for how the earth's rocks were formed.[9]

After Oxford, Lyell went to London to train as a barrister, becoming a member, as every barrister had to do, of one of the Inns of Court. He entered Lincoln's Inn in 1819 and was called to the bar, required to attend the courts as they moved from town to town in England.[10] But problems with his sight prevented him from continuing. His father wrote to a friend that Charles was nursing 'eyes which threaten to be permanently so weak & painful that the possibility of intense application & consequently of pursuing the law with effect becomes very doubtful . . . a temporary cessation from hard reading is indispensable . . .'[11] Love of geology would have pulled Lyell away in any case. In 1831 he complained to Gideon Mantell, a young geologist whom he had met ten years before on a field trip in Sussex, that he was 'buried in the study of law [and] I am too fond of geology to do both'.[12]

Lyell published his first scientific papers in 1825. The following year he became a fellow of the Royal Society and in the spring of 1831 he sought and won an appointment as professor of geology at the new (Anglican) King's College London (founded in 1829 to counterbalance that 'godless college in Gower Street', the new secular London University). King's being a clerical institution, the decision on Lyell had to be approved by no less than the Bishop of London, the Archbishop of Canterbury, the Bishop of Llandaff, and two doctors. The prelates, Lyell informed Mantell, 'considered some of my doctrines startling enough, but could not find that they were come by otherwise than in a straightforward manner . . . and that there was no reason to infer that I had made my theory from any hostile feeling towards revelation'.[13] (He was also honoured with an MA from Cambridge.) Fortunately for Lyell, King's seems not to have heard of his wish, expressed to the geologist Poulett Scrope, who was about to review *Principles*, 'to free the science from Moses'.[14]

The success of *Principles* – written in his lawyer's rooms in the Temple and Gray's Inn – sparked great interest in Lyell's forthcoming lectures. Some among Lyell's friends wished that ladies (mainly their wives and daughters) might come to hear them. But Lyell, in the misogynist tradition of his time, thought it would be 'unacademical' to admit women into the classroom.[15] In the end a few were admitted and the audience at his King's lectures swelled to nearly 300. The lectures were judged a great success, not least by Lyell himself. As he wrote to his fiancée, Mary Horner: 'I kept the attention of all fixed, by not reading, & you cd. have heard a pin drop when I paused.'[16]

Lyell felt he worked hard in his second lecture on the delicate subject of the relation of geology to natural theology, the branch of theology that attempts to prove the existence of God from natural observations. He had a good audience, including many of his friends and King's professors. He concluded with what he saw as a 'noble and eloquent passage' from the Bishop of London's inaugural discourse at King's, which proclaimed that 'Truth must always add to our admiration of the works of the creator [so] that one need never fear the result of free enquiry.'[17] Science, in short, could only glorify God. For this declaration, he was applauded.

Once appointed to King's, however, Lyell was disappointed to find that, as in Oxford, the geology lectures were extracurricular and therefore he could expect little income from fees for his lectures given on Tuesdays and Thursdays. For him this was a setback for, unlike many of the other early geologists, he was not wealthy and needed an income. When the publisher John Murray of Albemarle Street paid him the hefty sum of £400 (the equivalent of about £20,000 today) for the first edition of *Principles*, with a promise of further payment for future editions, Lyell saw that his future might lie in authorship. He resigned from King's, even though he had been there only two years.

In a journal entry intended for Mary Horner, Lyell spelled out his thoughts: 'If I could secure a handsome profit in my work, I should feel more free from all responsibility in cutting my cables at King's College. Do not think that my views in regard to science are taking a money-making, mercantile turn. What I want is, to secure the power of commanding time to advance my knowledge and fame, and at the same time to feel that in so doing I am not abandoning the interests of my family, and earning something more substantial than fame.'[18]

When the first volume of *Principles* sold out, Lyell knew he had made the right decision. He warmed to the compliment he received from Murray: 'There are very few authors, or ever have been, who could write profound science and make a book readable.'[19]

Principles would soon have a great influence on the young Charles Darwin, twelve years Lyell's junior (and a future close friend). The twenty-two-year-old Darwin, on the recommendation of his Cambridge tutor John Henslow, took the first volume of *Principles* with him as he set out in 1831 on his long voyage as naturalist on HMS *Beagle*. He set himself the task of reading it all before the ship reached its first stop.

Lyell's book alerted Darwin to the danger of 'undervaluing greatly the quantity of past time'.[20] It described, as if in anticipation of Darwin's visit in 1834, the five-foot elevation of part of the coast of Chile by an earthquake. When Darwin reached Chile, he witnessed for himself a devastating quake and tidal wave in Concepción.[21] But far more important was the effect that *Principles* had on Darwin's train of thought. It stirred him to wonder about the changes in life forms over time. He shifted his attention from rocks and fossils to man – a direction that would lead him to write *On the Origin of Species*, the world-changing book that appeared in 1859.

Lyell's *Principles* appeared in a Britain ready for new ideas. The bold Reform Bill of March 1832 had widened the voting franchise (even if only slightly) to adult male property owners. For the first time they were given the right to choose who should represent them. The bill created ninety-eight new seats and enabled fast-growing industrial cities such as Birmingham and Manchester to elect a Member of Parliament for the first time.

When Lyell's influential work first appeared, the age of the earth was accepted to be about 6,000 years. The Christian churches had ceased to take the Bible's 'Six Days of Creation' literally as six twenty-four-hour days; rather, they had expanded the 'days' to a thousand years each. Archbishop James Ussher of Armagh, in 1650, had methodically worked out the lifespans of the descendants of Adam. Combining these with his knowledge of the Hebrew calendar and other biblical records, the archbishop came up with the precise time of creation: the evening preceding Sunday, 23 October 4004 BC.[22] This date, widely accepted as fact, was printed in the margins of texts of the Book of Genesis.

From its start, geology challenged established religion. Excavations for coal mines and rail tunnels revealed fossils and ancient rocks which clearly had formed over more than 6,000 years, let alone in six days. The new science raised the tantalising question: did the biblical Flood have a geological basis? The Geological Society of London, founded in 1807, answered with a resounding 'No'. It refused to equate Noah's Flood with a universal deluge – much to the dismay of some religious believers within its ranks who had hoped that geology would confirm Genesis. As Jim Endersby, the Cambridge University historian of science, has expressed their dilemma: 'If, as the Bible claimed, this planet had been made as a habitation for humanity, why had its creator taken so long to get the tenants in?'[23]

In Britain, in particular, geology provoked a crisis of faith. While in France the Revolution had broken the grip of the Roman Catholic Church on the state, the Anglican Church of England had great political and intellectual influence on the expression of ideas. At Oxford and Cambridge, the professoriat were mainly beneficed clergy who received an ecclesiastical living from the Church. Moreover, as many as two-thirds of their students planned to take holy orders when they left university. What were they to preach about the place of humanity in nature?

Lyell began his *Principles* convinced that the process of geological change had been slower than anyone had imagined. It was now accepted that natural processes were manifestations of energy acting on or through matter. Volcanic eruptions were no longer seen as an expression of the anger of the gods of the underworld. Lyell, with his lawyer's training, knew how to make a good case. *Principles* introduced his readers to the new concept of an ancient earth. It went on to divide the 'agents of change' into the Aqueous (such as estuaries caused by the water of seas, rivers and tides) and the Igneous (the formation of mountains such as Vesuvius and Etna by a long series of volcanic eruptions over an immense period of time). Lyell freely acknowledged his debt to the Scottish physician and naturalist of the eighteenth century, James Hutton (1726–97).

In his two-volume work *Theory of the Earth*, Hutton declared that, after studying the continents and coasts of the earth, he found a balance between lands being created and destroyed.[24] Lyell, in his own book, called attention to the bold assertion Hutton had made before the newly formed Royal Society of Edinburgh, when he declared that he was attempting to explain changes in the earth's crust by natural agents. 'The ruins of an older world are visible in the structure of our planet,' Hutton said, 'and the strata which now compose our continent have been once beneath the sea, and were formed out of the waste of pre-existing continents.'[25] Hutton himself read the varied layers of rocks as evidence that the earth's age had to be measured in millions, not thousands, of years. He recognised that sediments had been laid down slowly at various times in the past, then heaved up and sometimes penetrated by igneous intrusions.

Hutton discovered what is now called 'deep time' on his Scottish acres. As a gentleman-farmer, as well as a naturalist, physician and chemist, Hutton had watched with dismay as the rains poured down on his Berwickshire fields, carrying away the soil and depositing it into streams. Why, Hutton wondered, had God created land only to destroy it?

Deeply devout, he answered his doubts with religious reasoning: the divinely ordered process was not only destroying land but creating it at the same time by washing down sediment from hills and mountains. In March and April 1785 he presented his theory of the earth as a system to the Royal Society of Edinburgh. He was more than an inquiring scientist; at times he could be a gifted phrase-maker. In the first volume of *Transactions of the Royal Society of Edinburgh* published three years later he created a stir with what became his most memorable declaration: that in his study of the earth he had found 'no vestige of a beginning, no prospect of an end' – a phrase that Lyell was to misquote in *Principles* as 'no traces of a beginning, no prospect of an end'.[26]

Hutton's Scotland was a fitting birthplace for British geology. The country's spectacular scenery forces the most casual observer to wonder how the rocks got that way. An extinct volcano, Arthur's Seat, looks down on the capital, Edinburgh, whose western border is formed by the dramatic Salisbury Crags. As a boy Hutton would have seen the large hollow in Arthur's Seat, created in 1744 by the great landslide which exposed a large piece of volcanic rock.[27]

Another of Hutton's assertions – that granite was a young rock – in itself challenged the biblical view of creation, which held that everything had been made at the same time. Looking for evidence to prove his theory, in September 1785 Hutton headed northeast from Edinburgh to the Grampian Hills and Glen Tilt to the confluence of two great rivers, the Dee and the Tay. There he found veins of red granite traversing black schist and primary limestone. The striking contrast of colours and textures showed that the granite had flowed molten into the limestone. In other words, the landscape had changed dramatically; the two kinds of rock had not appeared simultaneously. Hutton then took the pattern as evidence that the earth's age had to be measured in millions of years, rather than thousands.

Better proof was to be found not far away and Hutton sought it to answer his sceptics. In 1788, with two friends (both unconvinced by his ancient earth theory) and assisted by several farmhands, sixty-two-year-old Hutton headed in a boat along the southeast coast of Scotland to the cliffs of Siccar Point. There he made one of the most celebrated geological discoveries in history. The men did not need a single hammer to get the meaning of the rocks. They looked up at a dramatic promontory. At its base, vertical grey rocks stood in parallel, like a row of books. Overlying these lay flat layers of red sandstone. The two rock types of totally different composition in juxtaposition formed a classic picture of what geologists call an 'unconformity'.

To Hutton and his two learned companions – John Playfair, a forty-year-old professor of mathematics at the University of Edinburgh, and a wealthy twenty-seven-year-old geologist, Sir James Hall of Dunglass – it was glaringly obvious that the vertical grey slabs had once been flat sand on the sea floor. Over vast time these sediments had been turned into rock – literally petrified. Then earth forces had slowly tilted the whole mass onto its side (by crustal processes now understood to have been continental drift and subterranean heat) and new deposits had arrived on top, slowly solidifying into rock themselves.

The message of Siccar Point was inescapable. The three men were facing visible proof that Archbishop Ussher's calculation of the earth's age at around 6,000 years was wrong. Ludicrously wrong. Hundreds of millions of years had to have passed for the mass of Siccar Point to have reached the configuration they saw.

Today it is estimated that the grey lower layers of Siccar Point's rocks were deposited about 425 million years ago and the overlying red sandstone about 80 million years later. Deep time indeed.

Another scientist immersed in ascertaining the great age of the earth was the French mathematician, astronomer and physicist of pre-Revolutionary France, Georges-Louis Leclerc – known from 1725 as the Comte de Buffon. Buffon based his experiments on his knowledge that the earth's centre was hot (as those who dug mines knew very well). Heating two dozen small metal globes until they were glowing red, he then measured the rate at which these cooled to the point where he or his assistants could hold them in their hands. From this experiment he first placed the earth's age at 43,000 years, then amended it to 75,000 years.

For more than three decades Buffon had been preparing a comprehensive forty-four-volume *Histoire Naturelle*, of which three dozen volumes had been published by the time of his death in 1788. These and subsequent posthumous volumes encompassed his ambition to explain all of natural history, geology, anthropology and optics in a single encyclopaedic work. Into his massive text Buffon quietly inserted his startling view that the earth was many thousands of years old. He also allowed himself to venture that living species changed through time. In addition, he was the first to point out that different animals and plants were to be found in different parts of the world, citing the absence of what he called 'species identity' between the four-footed animals of North America and Europe.[28] Remarking on the similarities between humans and apes, he audaciously suggested they might have a common ancestry. For this bold idea Buffon has been declared a pioneer in the theory of inheritance of acquired characteristics.[29] Accused of heresy, Buffon formally recanted but quietly clung to his ideas.

Today, the eminent palaeontologist Richard Fortey believes that Buffon deserves admiration. 'Count Buffon may have got his estimate of the Earth's age based on its hypothetical cooling from the molten state entirely wrong,' he states, 'but he felt free to make an estimate without nodding to religious authorities or anyone else . . . assessment of time was part of a more general scepticism in the age of free enquiry. Britain was rather late on the scene.'[30]

In 1795 Hutton went on to publish his theory of the earth, declaring: 'With such wisdom has nature ordered things in the economy of this world, that the destruction of one continent is not brought about without the renovation of the earth in the production of another.'[31] He spelled out his conviction that the earth must be millions of years old in order to have produced Siccar Point. What had solidified the strata of loose materials on the ocean floor? Hutton's answer was 'the power of heat and operation of fusion' – a doctrine that is fairly accurate by today's knowledge, which places the temperature of earth's outer core as around 3,000 degrees Celsius.[32]

Had Hutton been able to write expository prose as well as, say, Lyell, he himself might have been recognised as geology's founder. But he wrote clumsily and chose to include in his book many unwieldy passages in French. Moreover, the obvious conflict with the Bible would cause Hutton to be reticent about his discovery. Having been called an atheist when he presented his ideas to the Royal Society of Edinburgh a few years earlier, he did not want to call undue attention to the glaring conflict of geology with Genesis.[33]

It was not until 1802 that Hutton's friend, John Playfair, a far better writer, penned *Illustrations of the Huttonian Theory of the Earth*, and made Hutton's discoveries accessible. (Lyell, preparing his *Principles*, relied heavily on Playfair's *Illustrations*.) In his own book, Playfair described seeing Siccar Point: 'We felt ourselves necessarily carried back to the time when the schistus on which we stood was yet at the bottom of the sea . . . The mind seemed to grow giddy by looking so far into the abyss of time.'[34]

In 1987, Hutton's eighteenth-century discovery was acclaimed by the eminent science writer, Stephen Jay Gould. In *Time's Arrow, Time's Cycle*, he wrote of Hutton: 'He burst the boundaries of time, thereby establishing geology's most distinctive and transforming contribution to human thought – Deep Time.'[35] The discovery, said Gould, imposed a 'great temporal limitation' upon human importance: 'the notion of an almost incomprehensible immensity, with human habitation restricted to a millimicrosecond at the very end!'[36]

Today, scientists estimate the age of the earth at roughly 4.6 billion years. The encompassing solar system is believed to have emerged around 13.7 billion years ago as a result of the 'Big Bang' – the collapse of a fragment of a giant molecular cloud. The earth, like other planets, was then formed by accretion from a rotating disc of dust and gas. Dense materials such as iron sank into the core. Lighter silicates and water rose near the surface. Four layers formed: inner core, outer core, mantle and crust – the inner core so hot that the outer core has remained molten. Most of the earth's mass lies in the mantle, a solid covering that can deform slowly in a plastic manner. The external crust is rocky and brittle, fracturing when the earth quakes.

By far the longest part of the earth's history is now understood to have been the time before any form of life began – an estimated 3.9 billion years ago. There were single cells in the ocean, and then, 540 million years ago, the first animals – creatures with heads, tails and segmented bodies – became diverse and abundant. This period is now referred to as the 'Cambrian explosion'.

Siccar Point, now a Scottish National Heritage site, has been called the world's most important geological site.[37] It remains today as Hutton and his companions saw it – a marker of the age of the earth. Acceptance of the evidence that mountains fell and rose again and that the human species was a relative newcomer to the planet had as profound an impact as did the ideas of Copernicus, Galileo, Newton, Darwin and Freud in their time. In 1687 Newton, whose *Principia Mathematica* (a title Lyell intentionally echoed in his own *Principles*) set out the mathematical principles of motion and gravity, never questioned the traditional narrative dating the earth's beginning at the creation. The seventeenth-century genius who opened so many secrets of motion and gravity remained blind to the antiquity of the earth.

2

HEALTHFUL EXERTION

In the 1830s geology was more than new: it was fashionable. The word 'geologist' was often preceded by the word 'gentleman'. Many drawings of early geologists at work make a top hat seem as essential as a hammer, even on the summit of Vesuvius. They also convey the message that these were mainly men of independent means.

The popularity of geology was boosted by its accessibility. Unlike older sciences such as astronomy or chemistry, any enthusiast could perform research. While many of the early British geologists were clergymen, others were scholars or writers who saw themselves as 'natural philosophers'. One such was George Eliot's partner, George Henry Lewes. Lewes was the author of books on physiology, animal life and philosophy and an early contributor to the new scientific journal *Nature*. Together the pair spent many days scouring the coastal rocks around Britain, collecting marine fossils, classifying them and storing them in glass jars. Lewes presented their findings in 1858 in a 414-page popular book, *Sea-side Studies at Ilfracombe, Tenby, the Scilly Isles, and Jersey*. In it he gave instructions on the proper equipment for embarking on a day's fossil-hunting:

> It is necessary to take with you from London, or any other large town, in or near which you may live, a geologist's hammer (let it be of reasonable size), and a cold chisel; to these add an oyster-knife, a paper-knife, a landing-net, and if your intentions are serious, a small crowbar. We now go to market for a basket. It must be tolerably large, and flat-bottomed. Having made that small

> investment, we turn into the chemist's and buy up all the wide-mouthed phials he will sell us – those used for quinine are the best; but as he probably will only have two or three to sell, we must take what we can get. The short squat bottles, with wooden caps, now sold for tooth-powder, are very convenient.[1]

Geology was also seen as health-giving. At a time when tuberculosis and myriad digestive ailments were rife, there was a constant search for healing waters and clear mountain air. Many were drawn to hammering rocks and hiking along steep trails because they felt the better for it. In 1817 William Fitton, a physician and an early fellow of the Geological Society of London, declared: 'Geology has this great advantage, of which not even Botany partakes more largely, that it leads continually to healthful and active exertion, amidst the grandest and most animating scenery of Nature.'[2]

Four years earlier, in his *Introduction to Geology*, Robert Bakewell, a professional surveyor who was not a member of the gentlemanly Geological Society, advanced health as 'an additional recommendation' to 'this useful and pleasing science': 'it leads its votaries to explore alpine districts, where the surrounding scenery and the salubrity of the air conspire to invigorate the health, and give to the mind a certain degree of elasticity and freshness, which will enable them on their return to discharge with greater alacrity the duties of active and social life'.[3]

Bakewell included some helpful information on the vegetable origin of coal, the fuel which had begun to be recognised as essential for the new developing industries. 'Coal,' he explained, 'comes from heaps of trees buried by inundations – under beds of clay, sand and gravel.' He went on to apologise for 'the irksome labour' of learning geological terms. He concluded his *Introduction to Geology* with a blatant self-advertisement. Addressed 'to Landed Proprietors', he wished to inform 'those noblemen and gentlemen who may honour this volume with their perusal, that he undertakes the mineralogical examination of estates, to ascertain the true nature and qualities of the soil, stone, and various minerals or metallic ores, and the uses to which they may be most profitably applied'.[4]

In 1831 an even grander accolade was awarded to geology by the eminent English astronomer and chemist Sir John Herschel in his *Preliminary Discourse on the Study of Natural Philosophy*: 'Geology, in the magnitude and sublimity of the objects of which it treats, undoubtedly ranks, in the scale of the sciences, next to astronomy.'[5]

Lyell, too, liked to compare the promise of the new science of geology to that of the older astronomy and looked forward with great optimism to the discoveries that geology would bring. As he wrote in *Principles*:

> Never, perhaps, did any science, with the exception of astronomy, unfold in an equally brief period, so many novel and unexpected truths, and overturn so many preconceived opinions. The senses had for ages declared the earth to be at rest, until the astronomer taught that it was carried through space with inconceivable rapidity. In like manner was the surface of this planet regarded as having remained unaltered since its creation until the geologist proved that it had been the theatre of reiterated change, and was still the subject of slow but never-ending fluctuations. The discovery of other systems in the boundless regions of space was the triumph of astronomy: to trace the same system through various transformations – to behold it at successive eras adorned with different hills and valleys, lakes and seas, and peopled with new inhabitants, was the delightful meed [reward] of geological research.[6]

He foresaw an exciting future for the new science of geology which calculated 'myriads of ages . . ., not by arithmetical computation, but by a train of physical events – a succession of phenomena in the animate and inanimate worlds – signs which convey to our minds more definite ideas than figures can do, of the immensity of time'.[7] What the long-term benefits of geology would be, he could not venture to say, but the practical advantages already derived were considerable and more would undoubtedly follow.

Geology rode the crest of the Romantic movement with its reverence for pastoral landscape and the beauties of Nature. This preoccupation may help to explain the subject's surge of popularity in Britain. As the scientific historian Roy Porter observes: 'spurred by Romanticism and muscular Christianity, nineteenth-century geologists celebrated "doing geology on your feet" as the hard-core activity of their science'.[8]

The poet-philosopher Samuel Taylor Coleridge (1771–1834), supported by a five-foot walking stick, conducted many epic walks through the Severn Valley and the Welsh hills, and had a passionate response to wild nature that, claims his biographer Richard Holmes, was so physical and direct that he felt almost at times like a child suckling at her rocky breasts: 'From Llanvunnog we walked over the mountains to Bala – most sublimely terrible!' wrote Coleridge. 'It was scorchingly hot – I applied my mouth ever and anon to the side of the Rocks and sucked in draughts of Water as cold as Ice.'[9] Humphry Davy, a friend of Coleridge who corrected the proofs of the second edition of *Lyrical Ballads*, used similar oral imagery writing in his own blank verse: 'For I have tasted of that sacred stream/Of science, whose delicious water flows, From Nature's bosom.'[10]

Yet not every Romantic poet appreciated the aesthetics of the new science. William Wordsworth recoiled from the spectacle of geologists thwacking the hills of his beloved Lake District. In *The Excursion*, written in 1814, the poet deplored:

He who with pocket-hammer smites the edge
Of luckless rock or prominent stone, disguised
In weather-stains or crusted o'er by Nature
With her first growths – detaching by the stroke
A chip or splinter – to resolve his doubts;

And, with that ready answer satisfied,
The substance classes by some barbarous name,
And hurries on . . .'[11]

The barbarous sound (to English ears) of geological terms such as 'greywacke', 'schist' and 'gneiss' derived from geology's German roots. Serious scientific study of the earth began as mineralogy on the Continent in the late eighteenth century for the purpose of assisting the growing mining industry in northern Europe. Schools of mines in Germany, France and Hungary analysed rocks, stones and minerals for their commercial value. France had an estimable *Journal des Mines* to publish its information. Other disciplines too – architecture, medicine and agriculture – sought information on the physical composition of the materials found in the earth's crust.

Leading this new field was Abraham Gottlob Werner. From 1775 Werner was professor of mineralogy at Freiburg School of Mines, the first mining school in Europe, where his charm and eloquence elevated his Bergakademie (mountain or hill academy) to the status of a great university. Lyell later described Werner as kindling 'enthusiasm in the minds of all his pupils, many of whom only intended at first to acquire a slight knowledge of mineralogy; but when they had once heard him, they devoted themselves to it as the business of their lives'.[12]

It was at Freiburg that geology became a scientific discipline, establishing some of the terminology that so annoyed Wordsworth and which, in some forms, still persists. Werner was the first to highlight the constant relations of certain mineral groups and their regular order of superposition. His native Saxony was an important mining centre, and it was here that the science of geology grew out of mineralogy and first became a scientific discipline. Werner called attention to the fact that the position of minerals in rocks was invaluable knowledge for the purposes of mining. So too was the grouping of rocks. He saw the economic use of minerals and their application to medicine.

He also observed the contribution of rocks to the soil, which itself influenced human wealth, intelligence, architecture and patterns of migration.

Werner had no doubt that the earth was very old. He observed that the lowest level of rocks, composed of granite, gneiss and schist, contained no fossils – that is, no remains of living things. He called these 'Primary'. The rocks above these he designated as 'Transition'. These included a type of dark gritty sandstone which he named 'greywacke' ('*Grau-wacke*' in German), characterised by embedded rock fragments and a few traces of life evidenced by a small number of fossils. Higher up still were layers of rocks he called 'Secondary' – sedimentary rocks, filled with fossils and derived either from the deposition of mineral and organic particles or from the breakdown of existing highly stratified rocks. On top of both levels lay the more recent 'Tertiary' rocks consisting of loose gravels, sand and clays.

In 1786 Werner expressed his conviction about the origin of the earth: 'The solid globe, insofar as we know it, was originally formed entirely from water.' He decided that the planet had been formed by multiple great deluges and that all its rocks had been precipitated from a common 'chaotic fluid' (hence the term 'Neptunism').

Lyell, criticising this theory in his *Principles of Geology* thirteen years after Werner's death, sarcastically pointed out that Werner was provincial: he had 'merely explored a small portion of Germany' yet took it as a prototype for the whole world and went on to teach that 'the whole surface of our planet, and all the mountain chains in the world, were made after the model of his own province'. Unfortunately, Lyell continued, 'the limited district examined by the Saxon professor was no type of the world, nor even of Europe' and his students were later to discover 'that "the master" had misinterpreted many of the appearances in the immediate neighbourhood of Freyberg'.[13]

By the start of the nineteenth century, the scientific centre of Europe had moved to Paris. French science, already strong, had benefited from the French Revolution, for in its wake new and revived institutions were created. These included the Ecole des Mines, re-established in 1794, the Institut National des Sciences et des Arts in 1795, the Muséum National d'Histoire Naturelle in 1793, and the Jardin des Plantes replacing the monarchistic Jardin du Roi.

Best known among the pioneering French naturalists was Georges (Baron) Cuvier. Born in 1769 to a bourgeois family in Montbéliard, then part of the German duchy of Württemberg, Cuvier spent the years of the Revolution as a tutor in Normandy. An ardent reader of books on natural history, he corresponded with leading naturalists of the time and at the age of twenty-seven was invited to lecture at the new Muséum National d'Histoire Naturelle, an institution much admired in Britain, which at the time had nothing comparable. Examining the skeletal remains of large mammals, Cuvier identified the mammoth, the mastodon and the megatherium (literally: huge beast), an elephant-size fur-covered creature like a giant sloth. He became professor of animal anatomy at the museum and remained there under Napoleon. In 1798 he asserted that there had been 'thousands of centuries' before man.

In 1812, Cuvier published his four-volume *Recherches sur les ossemens fossiles* ('Researches on fossil bones'), a blend of geology and comparative anatomy, in which he called attention to the significance of the relics of past forms of life – 'palaeontology', as the new science came to be called. The remains of extinct creatures in the Tertiary rocks found in the Paris Basin and also in private collections brought from Holland startled him by showing that large mammals such as mastodons, pterodactyls and elephants had lived in that region. To Cuvier's trained eye, the fossil teeth and jawbones showed them to have come from different species to those of modern Asian and African elephants. He could see that the climate had once been much warmer than in his day, that the geological record was marked by many dramatic breaks and that the sea had covered the land at many times.

Cuvier's work dazzled the novelist Honoré de Balzac, who in 1831 asked in *La Peau de chagrin* (*The Wild Ass's Skin*):

> Is not Cuvier the great poet of our era? Byron has given admirable expression to certain moral conflicts, but our immortal naturalist has reconstructed past worlds from a few bleached bones; has rebuilt cities, like Cadmus, with monsters' teeth; has animated forests with all the secrets of zoology gleaned from a piece of coal; has discovered a giant population from the footprints of a mammoth. These forms stand erect, grow large, and fill regions commensurate with their giant size . . .
>
> He can call up nothingness before you without the phrases of a charlatan. He searches a lump of gypsum, finds an impression in it, says to you, 'Behold!' All at once marble takes an animal shape, the dead come to life, the history of the world is laid open before you.[14]

Such Continental discoveries were of little immediate benefit to Britain, however. The Napoleonic Wars from 1799 to 1815 effectively confined the British to their island. There were notable exceptions. The newly ennobled Sir Humphry Davy, together with his assistant Michael Faraday, travelled to Paris in 1813 carrying a permit from Napoleon himself. Napoleon had presented Davy with a medal for his electrochemical work, leading to the London press criticising Davy for being unpatriotic in a time of war.[15] To a friend he countered: 'Some people say I ought not to accept this prize; and there have been foolish paragraphs in the papers to that effect; but if the two countries or governments are at war, the men of science are not.'[16]

For the most part, not only travel but the exchange of ideas was cut off. The exclusion was all the more deliberate because the British ruling classes were alarmed by the rise in Britain of a radical press whose books and pamphlets might bring about a replication of France's Revolution. So great was the fear of the French that in 1799 Herschel, as George III's astronomer, was secretly commissioned by the War Office to provide a hundred-guinea spy telescope to be mounted on the walls of Walmer Castle on the southeastern coast of Kent to give early warning of an approaching French invasion fleet.

The end of the wars with France would come as a much-anticipated liberation for British intellectuals and scientists. Writing from Suffolk in 1812, the Reverend Adam Sedgwick, a Cambridge University clergyman who would become one of the giants of British geology, described the intense excitement on hearing of victory at Salamanca:

> No railroads, and no telegrams then. So day by day we went out to meet the mail-coach on its first entrance . . . I said to myself, if England lose her freedom I will pack up all I have and go to settle along with my relations among the free-men of the United States. We had heard reports of good news, and I took my stand on a little hill that overlooks the London road along with my party. Several hundred of the inhabitants joined us. At length the mail-coach came in sight, rapidly nearing us. On its top was a sailor, waving the Union Jack over his head, and gaudy ribbons were streaming on all sides, the sure signs of victory.[17]

Peace, following the Duke of Wellington's victory at Waterloo in June 1815, put an end to Britain's isolation. There began a mass exchange of scholars across the Channel.

Some historians today use the term 'savants' for these pioneering academics, arguing that the word 'scientist' did not come into use until 1833. But the same early practitioners could very well have been called 'geologists' – the term having come into use in 1795. (The word 'geology' itself appeared in 1735 when *The Shorter Oxford*

English Dictionary described it as 'the science which treated the earth in general as well as investigating the earth's crust and the rock layers composing it'.)

Whatever their label, the scholars brought German and French knowledge of geology to a rapid industrialising Britain, while British geological adventurers were now free to wander abroad, hammer in hand, to inspect the crags and crevasses of their choice.

3

DOWN THE MINES

'Our civilisation . . . is founded on coal,' George Orwell wrote in 1937 in *The Road to Wigan Pier*. 'The machines that keep us alive, and the machines that make machines, are all directly or indirectly dependent upon coal. In the metabolism of the Western world the coal-miner is second in importance only to the man who ploughs the soil.' He then advised: 'When you go down a coal-mine it is important to try and get to the coal face where the "fillers" are at work . . . when the machines are roaring and the air is black with coal dust, and when you can actually see what the miners have to do. At those times, the place is like hell . . .'[1]

The rise of the new science of geology was driven by the urgent need to get coal out of the ground. Coal was a northern European fuel. Although the first-century Romans who lived in Britain had found it useful, they did not attempt mining; they simply dug up what they found lying near the surface.

Britain came late to the recognition of the importance of rock study, having lagged behind France, Germany, Hungary and Russia, where the art of mining had been taught in scientific institutions for decades (the St Petersburg School of Mines was founded as early as 1773). However, as the demand from its factories and steam engines grew, and mining technologies improved, Britain was able to quickly overtake its neighbours. By the 1830s Britain was producing more than half of the world's coal.

Deep-shaft mining began in Britain in the late eighteenth century and expanded rapidly. Industry developed techniques for excavating at greater and greater depths as well as a knowledge of where coal seams lay. Almost as important was to show where coal could not be found in order to save companies from drilling needless deep boreholes. Britain became pre-eminent in geological theory through Lyell's *Principles of Geology*, in which Lyell deplored Britain's slow start: 'Our miners have been left to themselves, almost without the assistance of scientific works in the English language, and without any "school of mines", to blunder their own way into a certain degree of practical skill.' This 'want of a system', he charged, was all the more deplorable in a country 'where so much capital is expended and often wasted, in mining adventures'.[2]

Many of the growing ranks of geologists in the 1820s were indifferent to theories of the earth's origin in fire or water. What mattered more were the great financial rewards waiting for those who could read the rocks and tell others where to dig. One important innovator was the geologist and Anglican parson William Conybeare. Conybeare's principal work, *Outlines of the Geology of England and Wales*, published in 1822 showed the location of the Carboniferous, coal-containing beds. Such precise information was of immense value to industry and intensified British interest in geology.

British coal proved so abundant that it was able to keep up with rising demand. During 1770–80 annual production output was about 6.25 million long (as opposed to metric) tons. Output soared after 1790 and reached 16 million tons by 1815, at the end of the Napoleonic Wars. Later, by 1881, British production had reached 184,300 million tons per annum, with nearly half a million people employed in mining.

It is hard now to appreciate that the famed industrial revolution was based entirely on an industry which sent men to work knowing that each day could be their last. By 1800 Britain's coal mines had men and boys working as deep as 1,000 feet below the surface. Wooden pit props were introduced to support the roofs in deep shafts, allowing easier access to even richer seams. Miners carried oil lamps or candles to light the darkness, risking the constant danger of igniting flammable gases.

Controlling the circulation of air and dangerous gases was vital. As early as 1815 the Sunderland Society for the Prevention of Accidents in Coalmines appealed to the most esteemed scientist in the land, Sir Humphry Davy, to find some safe way to illuminate the mines. His swift invention of the Davy Lamp made his already famous name even more famous. Protection for miners came with a lamp that achieved safety by restricting the entry of air to the wick lamp through a mesh screen. Yet the death rate did not drop dramatically: miners could be careless in its use, finding it a good way to light their tobacco pipes. That problem remained unsolved until the late 1880s when electric lighting was installed underground.

Still the underground tunnels carried the menace of flooding, collapse and explosions. At the all-too-frequent miners' funerals, male voice choirs, accompanied by colliery bands, sang moving songs such as 'Gresford'. The ballad was written soon after the Gresford Colliery disaster in northeast Wales in 1934, when around 250 miners were killed after an explosion of firedamp. Ventilation shafts collapsed and men and boys were entombed when the mine was sealed off to prevent further explosions. Despite desperate efforts to get the bodies out, only eleven were ever recovered. The reportorial song is still sung:

You've heard of the Gresford disaster,
The terrible price that was paid,
Two hundred and forty-two colliers were lost
And three men of a rescue brigade.

It occurred in the month of September,
At three in the morning, that pit
Was racked by a violent explosion
In the Dennis where gas lay so thick.

The gas in the Dennis deep section
Was packed there like snow in a drift,
And many a man had to leave the coalface
Before he had worked out his shift.
[. . .]

Down there in the dark they are lying,
They died for nine shillings a day.
They have worked out their shift and now they must lie
In the darkness until Judgement Day.

The Lord Mayor of London's collecting,
To help both our children and wives,
The owners have sent some white lilies
To pay for the poor colliers' lives.

Farewell, our dear wives and our children,
Farewell, our old comrades as well.
Don't send your sons down the dark dreary pit,
They'll be damned like the sinners in hell.

Other Victorian mining disasters included the Haswell explosion in Durham in 1844, which killed ninety-five colliers, and that at the Hartley Colliery in Durham in 1862, which took the lives of 204 men and boys.

The danger has persisted. Britain's mining archive lists more than 200,000 men, women and children as having suffered death or injury in the pursuit of coal up to the end of the twentieth century. In that century alone more than 100,000 coal miners were killed in the United States.

Yet a bigger fear in the second part of the nineteenth century was that the coal would run out. In 1865 a young economist, W. S. Jevons, wrote *The Coal Question* in which he warned that Britain's 'lavish use of cheap coal' could not continue. The coal might soon be exhausted and prosperity would disappear with it: 'We have to make the momentous choice between brief greatness and longer continued mediocrity.'[3] In fact, coal output rose and the price fell for many decades.

Even now, according to the science historian Matt Ridley, fossil fuel is not running out. He has called Jevons's gloomy warning 'hilariously timed, six years after the first oil wells were drilled in Pennsylvania'.[4]

4

VESTIGES OF PATERNITY

Geology had many fathers. Although no one ever claimed the title, a man to whom it was publicly accorded was the mapmaker William 'Strata' Smith.

The publication of Smith's 'Geological Map of England' in 1815 marked the beginning of geographical exploration in Britain. Working alone, and travelling much of the time on foot, Smith combed England, Wales and southern Scotland, examining the rocks and classifying the strata according to the fossils found in them. He discovered the same strata occurring in different patterns across the country. (He compared them to a pile of slices of bread and butter.[1]) Analysing these slices created the new art of stratigraphy.

Writing in the journal *British Critic* in 1831, Charles Lyell claimed that English success in the study of stratigraphy 'must make it hereafter appear one of the most remarkable passages in the history of science'. These early researches would lead to an understanding that stratigraphical classifications and configurations could be applied throughout the world.

In 1831, the Reverend Adam Sedgwick, president of the Geological Society of London, awarded Smith the society's first Wollaston Medal, saying it was fitting 'to place our first honours on the brow of the Father of English Geology'.[2] Smith was 'a great original discoverer in English Geology', he declared, 'and being the first, in this country, to discover and to teach the identification of strata, and to determine their succession, by means of their embedded fossils'.[3] The Wollaston Medal, made of gold and valued at ten guineas, was intended to form part of the annual award left by the bequest of William Wollaston, a chemist. Unfortunately for Smith, the new medal had not yet arrived from the Royal Mint. He had to be content with a purse containing twenty guineas and await the medal's later arrival. Even so, he made what the society felt was 'a short and manly speech'.[4] The Wollaston Medal remains geology's equivalent of a Nobel prize.

Sedgwick's accolade transformed Smith, a humble surveyor and orphaned son of a village blacksmith, from provincial folk hero into a major icon of British science. Smith had offered the first formulation of the law of strata identified by fossils – that is, the understanding that the proper sequence of rock strata can be ascertained by observing the fossils characteristic of each layer. In other words, two layers of rock from different sites could be regarded as of equal age if they contained the same fossils. Smith also opened the way to the understanding that the composition of the earth's crust was more economically useful and intellectually exciting than were theories of the forces by which that crust had been moulded.

Smith's brightly coloured map, published on 1 August 1815 and titled 'The Strata of England and Wales with part of Scotland', measured five feet by three feet and was printed across fifteen individual sheets. Showing the location of coal, chalk and low marshy ground or fens, the map was nothing less than a history of Britain told through the layers of its rocks. His great innovation was to trace rock formations or strata across country – hence his nickname 'Strata Smith'. One of his map's revelations was to show how the distinctive chalk of England's south coast (including the famed White Cliffs of Dover) forms an almost continuous outcrop reaching to the Midlands and northeast Yorkshire. (It was later understood that the deposition of the chalk occurred in what became known as the 'Cretaceous period', beginning about 140 million years ago.)

Smith's knowledge of Britain's geography came from the assignment he received from his employers, the Somerset Canal Company, who aimed to build a grand canal to serve the county's coal-mining industry. Smith was set the task of travelling the length and breadth of England and Wales to research how waterways were being constructed and linked together. For his tour, Smith travelled by coach, sitting up alongside the driver and a guard armed with a blunderbuss to fend off the highwaymen who were a constant menace.

In 1792 Smith climbed down into every one of the mines in the Somerset coalfield and saw what is today called 'the Westphalian stage of the Upper Carboniferous' – rocks laid down 304 to 312 million years ago. It was in the Mearns Colliery that Smith first noticed the succession of rock and fossil types that indicated their comparative ages. 'The stratification of the stones struck me as something very uncommon,' he wrote in a letter ('stratification' was a seventeenth-century word that would come into geological use in 1795 to refer to the order of the layers of sedimentary rock) 'and until I learned the technical terms of the strata,' Smith continued, 'and made a subterranean journey or two, I could not conceive a clear idea of what seemed so familiar to the colliers.'[5]

He then grasped that recognisable seams of coal would always be in the same relation to one another. His imagination went further. He saw it possible 'to map the underneath of England' – that is, to find and identify the outcrop of a particular stratum in one place, then find it in other maps and be able to 'extrapolate the position of that stratum as it snaked through the entire English underworld'.[6]

Smith also helped in plotting the routes of the canals themselves – the highways of the future. Plans were conceived for connecting the whole country – linking the Thames to the Mersey, for example – to carry not only coal but other products such as Wedgwood porcelains and barrels of ale. The alternative form of transport was primitive: horse-drawn wagons to haul coal from its sources to the factories and railways that depended on it, as by then did many homes for heating.

Smith's work is celebrated in the 2001 biography by Simon Winchester, *The Map That Changed the World*, which contends that Smith produced the first true geological map of any place in the world, one that in itself opened the way for making fortunes in oil, iron, coal, gold and diamonds.[7] (Winchester, while confessing that he would not want to dislodge his hero, Smith, from his pedestal as father of English geology, acknowledges that an earlier naturalist named John Rawthmell had noticed in the 1730s that the curious figured stones, not then recognised as fossils, occurred inside the rocks that stretched northeast from Dorset to Yorkshire.[8])

That one of the strongest candidates for geology's founding father should be British, owed much to the country's unique collection of rocks, representing almost every geological epoch, from the pre-Cambrian to the most recent. To have this history displayed over a relatively small space made travel and exploration comparatively easy. The convenience of geography can 'tempt the belief', wrote Winchester, 'that it is right and proper that the science of geology was born in Britain, and further adds emotional claim to William Smith being its most natural father'.[9]

The importance of Smith's map was recognised by the prime minister himself. Lord Liverpool, the Conservative leader, came to see it at Smith's modest London home off the Strand and personally congratulated him on a wonderful creation. Four hundred copies of the map were printed, numbered and signed, sold with the simple title: 'W. Smith's Discoveries of Regularities in the Strata'.

For Smith, such recognition may have helped to weaken his belief that geology was an upper-class occupation or, as he put it, 'the theory of geology was in the possession of one class of men, the practice in another'.[10] Certainly Lyell in his *Principles*, written fifteen years later, drew attention to Smith's humble background, describing him as 'an individual unassisted by the advantages of wealth or station in society', and calling his map 'a lasting monument of original talent and extraordinary perseverance, for he explored the whole country on foot without the guidance of previous observers, or the aid of fellow-labourers, and had succeeded into throwing into natural divisions the whole complicated series of British rocks'.[11] J. F. D'Aubuisson, whom Lyell described as 'a distinguished pupil of Werner', paid a just tribute to this remarkable performance, observing that 'what many celebrated mineralogists had only accomplished for a small part of Germany in the course of half a century, had been effected by a single individual for the whole of England'.[12]

Aside from Werner, a strong Continental candidate for geology's father is Georges Cuvier. Cuvier is now acknowledged as the scientist who brought zoology to bear on geology. He is unquestionably the father of vertebrate palaeontology – the study of fossil bones and the prehistoric life forms on earth. Cuvier maintained that anatomy must follow fundamental laws just as Newton's laws of physics do. If the fossil came from a carnivore, then it would display evidence of grasping forelimbs, sharp teeth, or strong hind limbs for catching and eating meat. From a single fossil bone (as Balzac had noticed), he could tell whether the beast was a mammal, a reptile or a bird.

With lively clear eyes, red hair and a strong mouth, Cuvier was an attractive man who had made a vivid impression during his visits to Britain after the Napoleonic Wars. He travelled to Oxford in 1818 in order to see William Buckland and the university's new museum display of recently uncovered dinosaur bones with huge teeth still attached to the jaw.

More than anyone, Cuvier was responsible for the remarkable change in geology that occurred by the end of the 1820s. His *Ossemens fossiles* rallied the first generation of English palaeontologists. As a great comparative anatomist who studied amphibians, reptiles, birds and mammals, he showed that the importance of fossils lay not just in their survival but in the clues they revealed as to how the parts had fitted together. The book's full title (*Recherches sur les ossemens fossiles de quadrupèdes, où l'on rétablit les caractères de plusiers espèces d'animaux que les révolutions du globe paroissent avoir détruites*) shows that Cuvier was as concerned with the disappearance of species as with their formation. He invited readers to follow him into the past and try to decipher and restore earth's history 'before any nations existed'.

Cuvier saw the poetry in geology. He envisioned that the new science might 'burst the limits of time' ('*franchir les limites du temps*') just as astronomy had burst the limits of space.[13] Cuvier's phrase accompanied a call to man to enjoy '*la gloire*' of reconstructing the history of thousands of centuries which preceded human existence.

Cuvier's major contribution was to pose the question raised by the ancient fossil bones. What had happened to make the species vanish? His own explanation was that great catastrophes had enveloped the planet and wiped out a number of species. That was the best theory he could offer four decades before the appearance of the idea of origin by evolution. But he had at least opened the debate about extinction of species. Until then the prevailing belief had been that the creatures identified by fossils must exist alive elsewhere in some undiscovered part of the world. The very idea of extinction challenged religious teaching. Why, as Hutton had wondered about land, had God created living creatures only to destroy them?

In 1796 Cuvier observed that the lower the stratum of rock, the greater the difference between its fossils and their present-day counterparts. He seems to have been the first to remark on this phenomenon. However, he did not recognise any progression from lower to higher as species developed. That would come later.

For the publishers John Murray, Charles Lyell was the 'father of modern geology'. So their archive claims. But Murray's would say that, wouldn't they? For the venerable publishing house, founded in 1768, *Principles* was one of their best sellers, alongside the works of Jane Austen, John Ruskin and Lord Byron. Nonetheless, Lyell is a strong contender for the title.

Like the great geologist James Hutton, Lyell was a Scot. Born in 1797 the eldest son of a Scottish landowner, Lyell was not, however, raised at his birthplace – the family's large estate at Kinnordy, in Forfarshire – but in the less geologically dramatic New Forest in Hampshire, where his parents lived at Bartley Lodge. His physical appearance was described by his good friend Gideon Mantell as unremarkable, except for 'a broad expanse of forehead . . . a decided Scottish physiognomy, small eyes, fine chin, and a rather proud or reserved expression of countenance'.[14]

Lyell's *Principles* raised geology to the status of a science by showing the constancy of physical laws throughout time. His fluency, imagination and breadth woke up an age to the importance of studying the earth's history. The volumes were widely read not only by scientists but by writers such as Dickens, George Eliot, John Stuart Mill and Henry Adams. The questions of species raised in the book opened the way to the evolutionary debate which would burst out with the publication of Darwin's *On the Origin of Species* in 1859. *Principles* was the first work in English (perhaps in any language) to try to present the earth's physical history. It carried also a philosophical message in its subtitle: *Being an Attempt to Explain the Former Changes of the Earth's Surface, by Reference to Causes Now in Operation*. Lyell proclaimed that all geological changes had taken place over millions of years through processes that continued to the present day. In this sense, Lyell may be regarded as a founding father of modern geology.

Lyell's book set flowing the full tide of what was called 'uniformitarianism'. This belief held that there had been no violent catastrophes, such as the convulsion that the ardent geologist William Conybeare maintained had thrown up the Alps in a single heave. Lyell believed that geological changes proceeded slowly. Rain and rivers wore down the land; volcanic action and earthquakes raised sediments from the bottom of the sea to form mountains. The division between 'uniformitarianism' and 'catastrophism' was defined by William Whewell, the philosopher of science, in 1832.

In the *Quarterly Review*, reviewing the second volume of *Principles*, Whewell asked whether, 'the changes which lead us from one geological state to another have been, on a long average, uniform in their intensity, or have they consisted of epochs of paroxysmal and catastrophic action, interposed between periods of comparative tranquillity? These two opinions will probably for some time divide the geological world into two sects, which may perhaps be designated as the *Uniformitarians* and the *Catastrophists*.'[15]

Aside from philosophising, Lyell laid down three practical rules for the geologist: 'travel, travel and travel' – advice which implied leisure, money and physical strength. Before he wrote his masterwork, Lyell practised what he would preach.

In 1827, as a young barrister, he visited Paris accompanied by the French scientist Alexandre Brongniart and Constant Prévost, one of Cuvier's former students. They took him to the Paris Basin and demonstrated what they believed was the geological history of the area: tranquil sediment that had lain there over a vast timescale that had been punctuated by occasional sudden alterations between marine and freshwater conditions. Prévost convinced Lyell that the evidence showed how the ancient environments paralleled those of the modern world – a powerful vindication of Lyell's faith that actual causes would be adequate to explain everything in the geohistorical record. Lyell was also introduced to Baron Cuvier himself.

Cuvier received guests in Paris every Saturday evening in the large drawing room of his library. Warmly welcoming Lyell, he invited him to call at his institute on Monday. Lyell obliged and brought with him a strange unidentified fossil tooth, found in the Tilgate Forest in Sussex, and given to him by Gideon Mantell. Cuvier declared that it was probably the upper incisor of a rhinoceros.[16]

The following year, once again turning himself from a barrister into a geologist, Lyell toured central France with his geologist friend and former army officer, Roderick Murchison, and Murchison's new wife, Charlotte. Lyell was interested in the extinct volcanoes around Auvergne known as *puys*, having read a study of them by the young English geologist George Scrope. Published in 1815, Scrope's *Considerations of Volcanos* implied that the landscape had been created step by step. Lyell went off to see for himself.

'Auvergne is beautiful,' wrote Lyell, 'rich wooded plains, picturesque towns, and the outline of the volcanic chain unlike any I ever saw.'[17] With innumerable old ruins to be sketched, lakes and waterfalls, he wondered that the English had not discovered the area. Everywhere he saw evidence that millions of years of rainfalls, rivers and eruptions had created the landscape. The sight convinced him that all geological changes took place over millions of years through processes that were still continuing. The fossil record, he decided, showed the relatively recent arrival of man. 'Although we have not yet ascertained the number of different periods at which the Alps gained accession to their height and width,' he wrote, 'yet we can affirm, that the last series of movements occurred when the seas were inhabited by *many existing species of animals*.'[18]

It was with eager anticipation of more travel, therefore, that he could inform readers of the distinguished *Quarterly Review*, for whom he was reviewing Scrope's *Memoir on the Geology of Central France*, that the Auvergne in the Massif Central was 'a theatre of extinct volcanos'. He reported that 'this fascinating area may be reached in a journey of less than forty hours by the public conveyance from Paris'.[19]

Lyell concluded that the rivers of the region, however small they appeared, had carved out their own valleys. When he reached Nice in August 1828, he found seashells 200 feet or more above the level of the sea. He realised at once that the shell-containing rocks must have been elevated since first deposited.[20]

When the Murchisons decided to return home to England, Lyell proceeded on his own to Italy. In Tuscany he worked out the geology while being driven in a gig on the road from Florence to Siena. To Rome he brought the same geological reporter's eye that would later enliven his *Travels in North America*. 'At Rome I found the geology of the city itself exceedingly interesting,' he wrote to his sister

Eleanor. 'The celebrated seven hills of which you have read, and which in fact are nine, are caused by the Tiber and some tributaries, which have cut open valleys almost entirely through volcanic ejected matter, covered by travertin containing lacustraine shells.'[21]

From Rome he went by post-chaise to Naples where, finding that his planned boat for Sicily would not leave for another twelve days, he took himself to the island of Ischia at the northern end of the Gulf of Naples. The journey was worthwhile. At an elevation of 2,600 feet he found the same kind of shellfish still living in the sea: another confirmation of uplift. Looking for signs of elevated beaches, once more he found them. He wrote to his sister that he would let the world know that Ischia had been populated with the same oysters and cockles 'which have now the honour of living with or being eaten by us'.[22]

Over the next few months, Lyell conducted an extended tour of Italy and Sicily. Near Naples he climbed Vesuvius, where he studied the lavas left after the great eruption six years earlier. In Sicily he ascended Etna, the largest volcano in Europe, where, on the summit, he was astounded to find once more what he had found in Ischia: fossil shells of the same species as were then living in the Mediterranean. These proved to him that Etna's summit had once been part of the sea floor. On the eastern side of Etna, he studied the dramatic circular crater called Valle de Bove and saw it as a cross-section of the mountain, revealing how the mountain had been built up slowly over an immense period of time. Etna, to him, was 'placed as if to give just & grand conceptions of Time to all in Europe'.[23] Around it were minor cones of widely different ages. 'Nothing can be more beautiful,' he wrote to Murchison, 'than the view from many parts of Etna down into these wooded volcanoes covered with oak & pines & with their craters variously shaped.'[24] His guide told him that Etna smoked most when the sea was high.

Lyell continued on a long circular tour of the eastern part of the island, ending up in Palermo on 29 December 1828, just before the rains came and turned the roads to mud. The trip committed him to his new chosen subject. 'I shall never hope to make money by geology, but not to lose, and tax others for my amusement,' he wrote to Murchison from Naples on 15 January 1829. 'My work is in part written and all planned. It will not pretend to give even an abstract of all that is known in geology, but it will endeavour to establish the *principle of reasoning* in the science; . . . that no causes whatever have, from the earliest time to which we can look back, to the present, ever acted, but those now acting; and that they never acted with different degrees of energy from that which they now exert.'[25]

Continuing his Italian travels, Lyell visited Pozzuoli, a port just to the west of Naples. (Lyell spelled it 'Puzzuoli'; its Latin name was 'Puteoli'.) The region was not the tourist attraction it would become after 1834 when Edward Bulwer-Lytton's vivid novel, *The Last Days of Pompeii*, depicted the people and animals of AD 79 trapped by the lava and ash spewed out by a cataclysmic eruption of Mount Vesuvius.

Pozzuoli's celebrated ancient monument, the Temple of Serapis, greeted Lyell's eyes like a vision from on high to tell him that his theory of earth movements was right: its three ancient columns, each over forty feet high, were riddled halfway up with mollusc holes. What plainer proof could there be that the ground on which the columns had been built had sunk below the sea and then been raised again? Clearly, while the pillars had been submerged, hard-shelled creatures had bored into the stone. Lyell could also recognise that the slow movements of lowering and raising had been gentle, for the columns had neither toppled nor cracked. He dated these movements to have happened 'since the Christian era'.[26]

Lyell was so pleased with this dramatic illustration of his theories that he used an etching of the 'Temple' as a frontispiece for the first volume of *Principles*.[27] Expanding on this evidence in *Principles*, he declared that there was scarcely any land in Europe, Northern Asia, or North America 'which has not been raised from the bosom of the deep';[28] if there were a new submergence, only the tops of the highest mountains would remain above the waters.

From other evidence gathered by the Geological Society, Lyell sketched a dramatic past for the British Isles as well. Animal remains found in coal and chalk showed that Britain's climate had once been much warmer and its land covered by a tropical sea. However, he warned that any who wished to comprehend the volcanic phenomenon must leave Britain and travel to countries where earthquakes were a frequent occurrence.[29]

His guiding tenet – that the present is the key to the past – was correct in essence. He stated that geological phenomena should be explained 'by reference to causes now in operation'. The shifting relations between land and sea, he said, were enough to produce the alterations in climate and land masses that had been observed. He reminded his readers that, despite the newness of geology, the laws of nature such as gravity held constant over time.

In his *Principles*, Lyell devoted an entire chapter to the work of the French aristocrat, Jean-Baptiste Pierre Antoine de Monet, the Chevalier de Lamarck. The 'father of the idea of progress in the development in forms of life', Lamarck was regarded by some as an 'upstart scientist'. His ideas remain controversial.

In 1793, Lamarck was appointed professor at the Jardin du Roi. In his *Philosophie Zoologique*, he argued that species did not become extinct but rather changed or transmuted into another, higher, form of creature. In his reasoning, wolves had become dogs. Species, he argued, by adapting to their environment moved upward and changed to higher forms of life. The neck of the giraffe had become longer by generations of stretching for higher and higher leaves. By 1801 Lamarck had classified spiders and crustaceans as distinct from insects. He had coined the word 'invertebrate' and argued his philosophy that there were no gaps in nature: one form of life grew out of another. In 1809 he published his evolutionary theory, arguing that all species had evolved in a continuous progression. Changes in structure arose from new conditions.

Lamarck had his later followers, notably twentieth-century Marxist biologists who found his views on the inheritance of acquired characteristics nicely compatible with Marxist teaching on the transferable effect of changes in society on future generations. Yet from the start, Lamarck's theories were easily disproved. The nineteenth-century German evolutionary biologist August Weismann cut off the tails of hundreds of rats over several generations and found that not one rat was subsequently born without a tail. Julian Huxley later gave a better example. By Lamarckian reasoning, he pointed out, Jewish male babies should be born without a foreskin as their fathers had none.[30] There was clearly no evidence that circumcision induced a genetically inherited characteristic.

In *Principles*, Lyell described at length Lamarck's thesis of transmutation of species. He then proceeded to try to demolish it in the following chapter, under the uncompromising title, 'Theory of the Transmutation of Species Untenable'.[31] He attributed the new preoccupation with species to the great varieties of animals and plants discovered in the previous half-century 'which poured in such multitudes into our museums'. He conceded 'none can doubt that there is a nearer approximation to a graduated scale of being'.[32] In his determined opposition to any idea of evolution, Lyell even abandoned the idea of Buckland, his Oxford professor, that there had been a traceable narrative of progress in the development of life. Instead he stressed the geographical causes of climate change and the piecemeal nature of the geological record that, to him, made it impossible to trace a clear path from the past to the present. In a letter to his wife, Mary, in October 1830, Lyell stated that if species of seashells showed no change after thousands of years, it 'must therefore have required a good time for Orang-Outangs to become men on Lamarckian principles'.[33]

Lyell's meandering thoughts reveal that he was utterly opposed to the idea that one species had transformed into another. 'An attack on evolution became central to the *Principles*,' writes geological historian James Secord, 'because of the threat Lamarck posed to the special status of humanity.'[34]

5

FIGHTING FELLOWS

British geologists shared their enthusiasm in the club they founded in a Covent Garden tavern on 13 November 1807. Briefly they considered calling their society 'mineralogical' but soon settled on 'geological': The Geological Society of London.

That the day was a Friday the thirteenth caused some merriment. Humphry Davy, an early member, joked that he never knew anything begun on a Friday the thirteenth to prosper. That the number of men present was also thirteen heightened the joke. The Freemasons' Tavern, where they met, occupied the front half of the spacious Freemasons' Hall on Great Queen Street off Drury Lane and was known for its good food. Sociability was the prime object. Members included doctors with an interest in chemicals and minerals – notably James Parkinson who, before turning his attention from medicine to geology, gave his name to the shaking palsy he was the first to describe. At that time doctors did consider themselves to be scientists; so did the Royal Society, which received a large number of academic papers from doctors.

The majority, however, were wealthy owners of land whose commercial potential they were eager to understand. Theirs was not a club for men with grubby fingernails; rather, a gentlemen's dining club. Within five years, the society included two dukes, two earls and ten clergymen. They recognised their common interest in sharing ideas, presenting their discoveries and adopting a common nomenclature (no small task) for the evidence they were accumulating with their hammers and sacks.

A constructive early decision was to disregard abstract theorising. The Geological Society would not fight the battle of the Neptunists who believed the world emerged from water, or the Plutonists who held that rocks were created by fire, and certainly not that of Genesis versus geology. Religion did not enter their debates in any way. Rather, they determined to be strictly empirical and to concentrate on collecting evidence – fossils in particular. These were men for whom a fine storage cabinet was a treasured possession and who were in awe of the Royal Institution's great collection, overseen by Davy, which held more than 3,000 minerals and fossils. They also reflected the general British fear of any speculative philosophising that might inspire a counterpart to the French Revolution.

The society's nickname – 'GeolSoc' (or 'Jollsoc' as it was and still is pronounced) – suggested a good time. It was an elite and expensive private club. The price of its dinner – fifteen shillings (about £60 today: a week's wages for many at that time) – in itself made it clear that the society was not for outdoor working men such as surveyors. The group grew rapidly. Within a year it had 136 members and a nomenclature committee 'to remove the confusion which now prevails'.[1] By 1810 it had eight committees, including one dedicated to chemical analysis, another to maps.

After only a year, the Geological Society boasted among its members thirty-seven fellows of the Royal Society. That distinguished society, chartered in 1662, which was and remains the most prestigious learned society in the world, acquired the 'Royal' designation in 1662 with the granting of a charter by Charles II. It initiated scientific publishing with its distinguished *Philosophical Transactions*, begun in 1665. (The seriousness of this journal may be judged by its subtitle: *Giving some Accompt of the Present Undertaking, Studies and Labours of the Ingenious in many Considerable Parts of the World*.)

The first president of the Geological Society was George Bellas Greenough, a well-travelled and wealthy bachelor of twenty-nine who had briefly been at Eton and Cambridge and who had seen not only Napoleon (in a visit to Paris during the peace of Amiens in 1802) but also the summits of Mont Blanc, Vesuvius and Etna. Greenough, who lived in a large Italianate mansion in Regent's Park not far from London Zoo, had just been made a fellow of the Royal Society as well as a Member of Parliament for the archetypal 'rotten borough' of Gatton in Surrey – a constituency notorious for having had at one time only one eligible voter. A former military man, he was a lawyer who had undertaken long geological tours with William Buckland in Britain and Ireland and who was a friend of Davy. He also had that geologist's essential: private means.

In 1819 Greenough published *A First Examination of the Principles of Geology* and then, in six sheets, his *Geological Map of England and Wales*, achieved with the help of the society's members who contributed details of rocks and strata and also with the results of his own long research begun as early as 1808. With more cartographic data than William Smith had been able to provide – and borrowing shamelessly from Smith (whose map he had seen in 1808 while it was in progress) – Greenough's map rapidly superseded Smith's in influence and sales. In 1865, after Smith's death, the Geological Society agreed that Greenough's map would in future be acknowledged as having been done 'by G. B. Greenough, Esq. FRS (on the basis of the original map of Wm. Smith, 1815)'.[2]

Indeed, many at the time felt that Greenough had stolen Smith's work. Greenough himself virtually acknowledged as much in a statement of apology for having appeared to be 'trespassing upon ground which I knew of right of pre-occupancy, his'.[3] The two maps resembled each other, he said, because both were correct and 'it is impossible that the views, the opportunities and the reasonings of two persons engaged on the same subject should be invariably the same'.[4] He sent Smith a copy of his map, which reached him in

Yorkshire where he had retreated to escape from his accumulated troubles – money woes and a wife who was losing her mind.

Another reason why the Geological Society looked at Smith with some suspicion (according to the society's historian Hugh Torrens) was that Smith was a theoretician. Smith's map was based firmly on his personal opinion that each of the strata of the earth's crust was characterised by its own distinctive fossils and by their ordered sequence. This conviction led to the kind of theorising and controversy that the new society was trying to avoid.

Smith would never be invited to join the Geological Society: the group was entirely out of his class. Later, when the society moved to new premises at Somerset House, Roderick Murchison, from his perspective as a member and wealthy landowner, found himself looking down on 'many of the Johnny Raws who come to SH'.[5] Even if he had considered joining, William Smith would have found the initial admission fee of six guineas prohibitive, not to mention the annual contribution of three guineas. He was beset with debt. Selling his geological collection to the British Museum in two lots – in 1816 and 1818 – did not save him. In 1819 he was committed to King's Bench Prison in Southwark for nearly ten weeks. The Geological Society did nothing to help him – in contrast to its gift of £1,000 to the wealthy Greenough, to enable him to complete his own map, which went on to sell very well.

A distinguished early member of the GeolSoc was the president of the Royal Society himself, Sir Joseph Banks, who was also director of the Royal Botanic Gardens at Kew. Joining the new geological group, Banks tried to persuade it to become a subordinate of the Royal Society. He failed. When after a year he saw the Geological Society taking on the aspect of an autonomous learned society, with a library and collections of its own, he confronted Greenough head on:

BANKS: So you intend to withdraw yourself wholly from the Royal Society, do you?
GREENOUGH: How so Sir Joseph?
BANKS: In regard to papers.
GREENOUGH: So far from it.[6]

Banks's concern was that the Geological Society would attract papers which would otherwise have been presented to the Royal Society and published in its *Transactions*. Greenough assured Banks that should the Geological Society ever receive a paper the Royal Society wanted, the Royal would have prior claim and his own society would hand the paper over to the senior society for publication in its own journal.

Unpersuaded, Banks immediately resigned from the geological upstart. So did Davy and two other FRSes. The Royal Society was supreme. Such was its reputation that when Davy was in Paris in 1813 with Napoleon's encouragement, at a dinner hosted in his honour by the Société Philosophique, a toast was drunk to the Royal Society rather than to the King of England. Greenough himself, however, did not resign from the geologists' club. He and some others managed to play both sides of the street – retaining memberships in both the Royal and Geological societies, Greenough remaining a Member of Parliament as well.

Banks's anxiety was not misplaced. The growing interest in fossil bones was shifting from the older to the newer scientific society. The geologists themselves were changing their focus as the younger members who joined were more interested in the newer fossiliferous rocks than the ancient hard rocks that held no trace of life.

The Geological Society was known for its lively and heated arguments, quite unlike anything ever seen at the staid Royal Society. There, after a paper was read, all would sit in dignified silence. At the GeolSoc, on the other hand, after someone presented a paper, boisterous arguments would often break out – usually but not always good-tempered. The meetings provided a spectator sport, with members bringing guests to watch the fun. The society met twice a month in a room lit by gaslight (another technical innovation) with reptile jawbones and skulls spread out on the table around which the listeners sat on benches as if in the House of Commons.

The parliamentary comparison struck the mathematician and philosopher Charles Babbage (originator of the concept of the computer), who was a frequent guest. He wrote that the meetings possessed 'all the freshness, the vigour, and the ardour of youth in the pursuit of a youthful science'. To him, the society had succeeded 'in its unusual experiment of having an oral discussion of the papers read at its meetings. To say of these discussions that they are very entertaining is the least part of the praise which is due to them.'[7] His compliment was outdone by J. S. Lockhart, the editor of the new *Quarterly Review*, who came along to enjoy the spectacle. 'Though I don't care much for geology,' Lockhart famously observed, 'I do like to see the fellows fight.'[8] Following the discussions, or fights, all dined together.

The Geological Society continued to grow, reaching a membership of 400 by 1818. In 1811 the society acquired its most amusing and soon-to-be most influential member, William Buckland of Oxford. It was he who in 1819 introduced Charles Lyell, who had been one of his students, to the society.

In step with the work of Cuvier in Paris, the Geological Society gradually shifted its principal focus from rocks to fossils. Soon the society adopted a constitution, then set up a library, and most important to its members, established a collection of rocks and fossils. Dr William Babington, one of the physicians among them, gave them a cabinet in which to hold the treasures. The society's popularity required a move to Bedford Street, north of the Strand, where the society remained from 1816 to 1828.

It was at Bedford Street, on 20 February 1824, that a dramatic occurrence took place. The young Reverend William Conybeare, a distinguished early geologist and fossil analyst, planned to discuss a near-complete skeleton of a huge marine reptile that had been recently dug out of the rocks at Lyme Regis in Devon. Conybeare was then rector of Sully in South Wales but lived across the Severn in Bristol, and started work on the extinct marine reptiles along local shorelines. For his important investigation of these fossils he was elected to the Royal Society in 1819.

The dramatic skeleton about which he was to speak had been discovered by the fossil-hunter Mary Anning on the evening of 10 December 1823; but it was Conybeare who gave it the Latin tag *plesiosaur*, meaning 'near to reptile'. The name stuck, as Conybeare was acknowledged master of knowledge about the ancient creature. The skeleton had an extraordinarily long neck – thirty-five vertebrae, unlike the usual three to eight.

In Paris, Cuvier heard a description of the find and did not believe it. He wrote to Conybeare suggesting that the skeleton might be a hoax (a phenomenon not unknown to fossil collectors at the time). The great French comparative anatomist could not accept the possibility of a reptile with thirty-five vertebrae in its neck. Conybeare was enraged.

Buckland, about to assume the presidency of the Geological Society, persuaded Conybeare to ship the long-necked reptile skeleton to London and to discuss it at the society's next meeting. But shipping was no simple task. Conybeare reached London before his fossil did. His presentation had to be delayed because the skeleton, embedded in a huge slab of rock, was on board a vessel which became stuck in the English Channel for ten days. When at last the huge package arrived at the society's headquarters, he faced the task of carrying it upstairs to the first-floor meeting room. With ten workmen, he spent a whole day trying to haul the slab up, but without success.[9] The bony trophy was left to rest in a dingy passage outside and the geologists were forced to peer at it by candlelight.

Conybeare at least had the satisfaction of delivering to the meeting his paper, 'Notice on the Discovery of an Almost Perfect Skeleton of the Plesiosaurus'.[10] A good audience had assembled in anticipation of the great news about to be revealed. Gideon Mantell was there – appropriately as the bones about to be described were from Dorest, where he had made his own giant bone discovery which he would present to the society in a few months' time. His friend Lyell was there too, with two guests.

No one was disappointed. Conybeare described 'the most monstrous creature ever discovered' and explained that, thanks to its excellent preservation in the chalk for thousands of years, the bones were almost in their original state. 'To the head of the Lizard, it united the teeth of the Crocodile, a neck of enormous length, resembling the body of a Serpent; a trunk and tail having the proportions of an ordinary quadruped, the ribs of a Chameleon, and the paddles of a Whale.'[11] There were ninety joints in the backbone (the large number of vertebrae which had aroused Cuvier's scepticism) – greater than that of any other animal. 'That it was aquatic,' Conybeare declared, 'is evident from the form of the paddles.'[12]

Conybeare had written his paper in collaboration with Henry De la Beche, a geologist and brilliant draughtsman who contributed a beautiful anatomically detailed sketch of the plesiosaur's head. The fossil's bones had been found flattened, but De la Beche's sketch reconstituted the head with its glaring eye socket; it also alphabetised the many components. The powerful drawing made sure that Conybeare's discovery was known about in Paris.

De la Beche himself could not attend the presentation. However, the graphic words Conybeare used in his talk are illustrated in the sketch that became De la Beche's most famous artwork – *Duria Antiquior* ('More Ancient Dorset'). The plesiosaur's head, said Conybeare, was 'remarkably small' compared to that of the *Ichthyosaurus* itself: 'its long neck must have impeded its progress through the water; presenting a striking contrast to the organisation which so admirably fits the *Ichthyosaurus* to cut through the waves'. Thus it 'swam upon or near the surface, arching back its long neck like the swan, and occasionally darting it down at the fish which happened to float within its reach. It may have lurked in shoal water along the coast, concealed among the sea weed, finding a secure retreat from the assaults of dangerous enemies.'[13]

Next, Conybeare handed the floor to the flamboyant William Buckland, who was making his first address as president and was waiting to announce a great discovery of his own. In fact, Buckland's beast was even greater: a *Megalosaurus* – now known to have been a dinosaur. Buckland did not make too little of it, recreating for the members an imagined scene, which involved half-starved scavenging hyenas emerging from caves to help themselves to water rats in a lake nearby.

Lyell was discomfited, confiding to Mantell: 'Buckland in his usual style enlarged on the marvel with such a strange mixture of the humorous and the serious, that we cd. none of us discern how far he believed himself what he said.'[14] Conybeare wrote to reassure De la Beche that their plesiosaur presentation had not come up short: 'I made my beast roar almost as loud as Buckland's hyenas.'[15]

By 1825 the Geological Society became royal in all but name when King George IV granted it a royal charter. At that point the society might have appended the 'royal' prefix to its name. However, in its eagerness to avoid offending the great Royal Society, and to discourage further resignations from its ranks or – a more alarming possibility – a complaint to the Privy Council, which would have involved great expense, the GeolSoc decided not to use the grand adjective. Even so, its members called themselves 'fellows' and were entitled to put after their names the initials FGS (Fellow of the Geological Society).

The purpose of the society, as George IV recognised (or was told to recognise), was 'Investigating the Mineral Structure of the Earth'.[16] That objective was quite different from exploration of the globe. Exploring was the province of geography, a mission soon to be undertaken by the Royal Geographical Society, founded five years later in 1830, with a strong emphasis on expeditions and discovery in the countries of the British Empire.

By 1824 the Geological Society had a new Geological Dining Club, the original one having lapsed. At the first meeting, thirty fellows were listed of whom twenty-four were members also of another club, the Athenaeum, formed in 1824 and located nearby. The Dining Club's members were limited to forty – a group which took over the de facto running of the society.

By then the Geological and the Royal Society were neighbours. In 1828, thanks to the intervention of the home secretary, Sir Robert Peel, who secured the consent of the lords commissioners of the Treasury, the Geological Society was allowed to move to Somerset House on the Strand, into rooms offered rent free by the Royal Society, as it did not need them. The GeolSoc's new premises overlooked the new Waterloo Bridge opened in 18 June 1817, at the western end of Somerset House. The house, an imposing structure made of granite, had three great arches where boats and barges could land. (In the early years of Somerset House, before the Thames Embankment was built, water lapped upon its south wing.) The society's rooms were duly refurbished by the neo-classical architect Decimus Burton (a GeolSoc member) at a cost of £683.4s.1½d. (over £50,000 today). Lyell pronounced himself very pleased with what he saw as 'our magnificent apartment'.[17] The first meeting was held there on 7 November 1828.

The glamour of geology, brightened by its increasingly distinguished members, made it feel a social step up simply to be on the GeolSoc roster. In time it came to include Thomas Arnold, headmaster of Rugby School; Sir Robert Peel when he was later prime minister; the King of the Belgians, the Crown Prince of Denmark and the Archduke of Austria.

Women were not allowed to be members of GeolSoc. It was hardly an issue, as no one considered that they should be. Not until 1870 would they be admitted to British universities and not to London gentlemen's clubs until well into the twentieth century. That women of the early nineteenth century were clearly interested in science is proven by photographs taken at the Royal Institution lectures which show a great number of women present. When the British Association for the Advancement of Science, formed in 1831, rejected the idea of women as members, Buckland advised Murchison: 'Everyone agrees that, if the meeting is to be of scientific utility, ladies ought not to attend the reading of papers and especially at Oxford as it would at once turn the thing into a sort of Albemarle-dilettante-meeting, instead of a serious philosophical union of working men.'[18]

Among the GeolSoc's members, several – notably Roderick Murchison and Charles Lyell – drew great benefit from well-educated childless wives who liked geology and were good at classifying, sketching, labelling specimens, and occasionally translating. In addition, Charlotte Murchison (unkindly described by Benjamin Disraeli as the 'silent wife' of 'a stiff geological prig') inherited great wealth which enabled the Murchisons to establish their intellectual *salon* in Belgravia – 'one of the hospitable scientific centres of London', according to Geikie. Mary Horner, meanwhile, had waited patiently through a year-long engagement to Charles Lyell, supporting his theories and writing as he struggled to finish his masterpiece. He had met her through her father, the Scottish geologist Leonard Horner, another GeolSoc member. Mary was an accomplished linguist and helped Charles in his travels, not least by sketching and classifying the specimens he found. After marrying in July 1832 they enjoyed a geological honeymoon, touring Germany, Italy, Switzerland and France – the first of many geological voyages they would take together. They would have no children – a fact that may have helped them form such an effective working partnership.

One of the most heated arguments involving the Geological Society – later known as 'the Great Devonian Controversy' – erupted in 1834. It arose over plant fossils found in Devon in a lower stratum than those in which fossils were usually observed. The discoverer was Henry De la Beche, who was preparing, for a fee, a geological survey of Devon for the Ordnance Survey. (Formed in 1791 to map Great Britain on a systematic scale, this group later became the British Geological Survey, a research and advisory service still active today.) While surveying, De la Beche had found plant fossils within a stratum far below the Carboniferous coal-containing formations. The leafy patterns lay within the gritty greywacke in Werner's German designation, which was classified as a 'Transition' rock. To accompany his contention, De la Beche collected the best specimens he could, but they were poorly preserved. They lay on the table as his note was read out at the GeolSoc. An impassioned discussion began.

His claim was no trivial matter. Geology had been increasingly successful in identifying coal seams (or 'coal measures' as they were called from an old miners' term), which became more and more essential for industries at home and abroad.

De la Beche's letter was held back until the end of the GeolSoc meeting, which had drawn eighteen members and twenty-seven guests in anticipation of the outburst to come. Roderick Murchison, a former army officer who had not lost his military bearing, led the attack. He was furious, scathing and condescending – his vitriol all the nastier for its snobbery. It was known to all present that De la Beche,* once wealthy with an income of about £3,000 a year from

*De la Beche pronounced his surname 'Beach', his father having embellished the family's original name 'Beach' by changing it to 'De la Beche', so as to echo that of a medieval Baron De la Beche of Aldworth who died in 1345; yet he did not change the original pronunciation.

estates in Jamaica, had lost his fortune in 1831 after an uprising of slaves. The loss imperilled his whole career as a gentleman of science. He had been compelled to resign his post as secretary of the Geological Society and to apply to the government for a grant to complete his ordnance survey of Devon.[19] Now, without a fortune, he was seen by his fellow geologists as a 'jobber' – that is, not a gentleman, but rather a man of a lower class who had to work for a living. Indeed, he was so strapped for money that he could not even afford the coach fare to London to present his greywacke case in person.

Murchison, supported by Lyell, insisted that the true greywacke of Devon could not possibly contain the fossil plants De la Beche claimed to have found within it. Both men freely acknowledged they had never been to north Devon. Even so, Murchison rejected De la Beche's claim that the rocks on the table were greywacke or Transition rocks. He was astonished, he said, that so experienced a geologist should have fallen into so great a mistake. There were no land plants before the Transition period – the ancient period when few traces of life could be detected.

The controversy would boil on for years. It sprang, although unrecognised by the participants at the time, from the disagreement and confusion over the new and important question: should rocks be dated by rock type or by fossil content? Today the answer is known to be both.

In one talent De la Beche was unquestionably supreme. He drew brilliant caricatures without which the history of geology would be the poorer. His 1830 masterpiece *Duria Antiquior* showed prehistoric creatures – the plesiosaur, the ichthyosaur and the pterodactyl – in the air, on land and below the sea, flying, swimming, defecating, eating each other. Below the terrified plesiosaur being bitten by an ichthyosaur, a pile of droppings (coprolites) is accumulating. The picture is considered to have been the first scene out of 'deep time' to be recreated – possibly the first attempt by the human imagination to portray what earth was like before man walked on it. Subsequently, through its copies, widely circulated, *Duria Antiquior* created a genre, as Tom Sharpe of the National Museum of Wales has pointed out, that has led to today's computer-generated reconstructions such as the film *Jurassic Park* and the BBC's *Walking with Dinosaurs*.

De la Beche's cartoon of the Great Devonian Controversy is a brilliant sideswipe at his GeolSoc doubters. It shows him, in a practical topcoat suitable for fieldwork, facing his critics, who wear elegant tailcoats. They hold lorgnettes up to their eyes as they stare at the cringing De la Beche, who, pointing to his nose, declares: 'This, Gentlemen, is my Nose', to which they respond: 'My dear fellow – your account of yourself generally may be very well but as we have classed you, before we saw you, among men without noses, you cannot possibly have a nose.' Gifted and controversial, De la Beche was a colourful figure, with a fondness for bright checked waistcoats and gold-rimmed spectacles. His glasses identify him in the classic early sketch of the GeolSoc sitting around its T-shaped table.

Along with his artistic gifts, De la Beche had an outstanding capacity for administration. In 1835 he was appointed superintendent of the small Museum of Economic Geology in Craig's Court near Charing Cross, an institution formed to show the application of geology to the useful purposes of life. Among his duties was to sit on a commission in 1838 to select the building stone for the new houses of Parliament. (They chose a sand-coloured Yorkshire limestone.)

In the end, the science of geology was helped by the fight, furious though it was, and De la Beche could be said to have pioneered the career of the professional geologist, transforming what had been a pastime for the privileged few into a serious career opportunity for anyone with the passion and skill to pursue it.

6

DATING THE DELUGE

If not the father of geology, the Reverend William Buckland was its best publicist. In 1818, when thirty-two and a reader of mineralogy at Oxford University (a post equivalent to a professorship, yielding £100 per annum plus £100 for lectures), he persuaded the prince regent to endow a readership in geology as well and to give him the title he craved: 'Reader of Geology'. His enthusiasm was not shared by other Oxford dons. When Buckland went on a tour of Italy, Dean Gaisford exclaimed: 'Well, Buckland has gone to Italy; so, thank God, we shall hear no more of this geology.'[1]

The natural sciences were not developing at Oxford under a utilitarian guise. They were studied neither for economic nor financial reasons, nor for the purpose of professional training. Quite the contrary. It was the perception of geology as history (albeit history of an antediluvian world) as a branch of humanistic learning that made the subject suitable as a fashionable branch of liberal education. Buckland explicitly stated in his inaugural lecture in 1819 that geology should be given a place at Oxford because it was 'founded upon other and nobler views than those of mere pecuniary profit and tangible advantage. The human mind has an appetite for truth of every kind, physical as well as Moral; and the real utility of Science is to afford gratification to this appetite.'[2]

Both Oxford and Cambridge had been under attack by the *Edinburgh Review* and other periodicals for neglect of secular learning. Oxford was particularly slow to develop the applied sciences such as medicine and agriculture, failing to keep pace with its older rival, Cambridge, by insisting on much more Greek. Cambridge, where mathematics held a prominent place in the examination system, already had a strong commitment to giving its students a knowledge of science. Neither university offered science as training for professional scientists. As chemistry, botany and geology were extracurricular in the first half of the nineteenth century, there were no examinations in these subjects. Science was offered as an option for Christian gentlemen, half of whom would become clergymen. As James Secord has observed, a theology of nature was one way of maintaining the authority of the state Church.[3]

Buckland, who was ordained a priest and made a fellow of Corpus Christi College, Oxford, in 1808, presented the new scientific subjects as complementary to the university's system of classical education. In his inaugural lecture, '*Vindiciae geologicae*', he respectfully referred to their introduction as an 'ingrafting (if I may so call it) of the study of the new and curious sciences of Geology and Mineralogy, on that ancient and venerable stock of classical literature from which the English system has imparted to its followers a refinement of taste peculiarly their own'. In Buckland's opinion, the new sciences were a necessary part of a proper liberal education and should be 'admitted to serve at least a subordinate ministry in the temple of our Academical Institutions'.[4]

'*Vindiciae geologicae*' was audaciously subtitled: 'or the Connexion of Geology with Religion Explained', making him the pioneer in the field which still thrives today: reconciling Genesis and geology. His address, both defensive and aggressive, tackled the crucial question head on: why should a university devoted to classics and training clergy include geology in its curriculum?

Buckland's answer was just what the fledgling clerics needed to hear. The science of geology, he asserted, gave historical support for Holy Writ and was consistent with the biblical account of creation and Noah's Flood. In his words: 'Again, the grand fact of a universal deluge at no very remote period is proved on grounds so decisive and incontrovertible that, had we never heard of such an event from Scripture, or any other authority, Geology of itself must have called in the assistance of some such catastrophe, to explain the phenomena which are unintelligible without recourse to a deluge exerting its ravages at a period not more ancient than that announced in the Book of Genesis.'[5]

Confidently he declared the 'beginning' in Genesis to be metaphoric; the only mental exercise needed was to interpret the word as referring to the timeless interval between the creation of the earth and man. In any case, he maintained, the very order in nature was proof of 'the supreme intelligent author' of creation.[6] It was not surprising, therefore, that 'the real utility of Science is to afford gratification to [this] large and rational species of curiosity'.[7]

The son of a clergyman, Buckland was born in 1784 in Axminster, Dorset, six miles from the beautiful ninety-five-mile-long shoreline formed by rocks laid down in the Jurassic period (about 200 to 150 million years ago) when shallow seas, far higher than today's, covered much of what is now Western Europe. (This famed stretch from west Dorset to east Devon, known as the Jurassic Coast, is now a UNESCO World Heritage Site. Its remarkable high limestone cliffs full of marine sediments show interleaved horizontal bands of shale and blue limestone, known as the 'Blue Lias'.) He became interested in fossils when walking with his father, the Rector of Templeton and Trusham, around the quarries and cliffs. Of these rocks, he later told his daughter: 'They were my geological school. They stared me in the face, they wooed me and caressed me, saying at every turn, Pray, Pray, be a geologist!'[8] At first, he had called his new passion 'undergroundology', but soon dropped the neologism for 'geology'.

With his education guided by an uncle (his father having become blind), Buckland went to school at St Mary's College, Winchester, and in 1801 won a scholarship to study for the ministry at Corpus Christi, Oxford. It was then that he began stuffing rooms with bones and fossils. He received his BA degree in classics and theology in 1804, having supplemented his scholarship funds by tutoring pupils in classics. He attended the lectures of regius professor John Kidd and benefited from the acquaintance of a younger student, William Broderip. Broderip was working on the succession of strata and was an expert conchologist – student of shells – who opened Buckland's mind to the importance of their gradations. Together they would walk Shotover Hill, east of Oxford, Broderip pointing out fossil shells which, Buckland later said, 'formed the nucleus of my collection for my own cabinet'.[9] From then on, he made many excursions on horseback, often wearing a white cloak, riding an old mare and carrying his heavy blue bag of fossils and hammers.

Buckland had been greatly assisted in preparing his inaugural lecture by his colleague, Reverend William Conybeare, much of whose energies were expended on making geology acceptable to Anglicans. Conybeare, aside from being a cleric (he was made Bishop of Llandaff in 1848) was also one of the most experienced geologists of his time. An early member of the GeolSoc, he had joined in 1809 as the leading authority on the fossil reptiles he had named ichthyosaurs and plesiosaurs. He was already persuaded that the position of fossils in rock layers showed their age. At Oxford, Conybeare became a close friend of Buckland, three years his senior. The pair went to the north coast of Ireland together and published a description of one of its features, the Giant's Causeway, in *Transactions of the Geological Society*.[10]

In the summer of 1811, Conybeare made a long geological expedition to southwest Wales with John Kidd, whose lectures, delivered in the basement of the Ashmolean Museum, he had much enjoyed. When in 1813 Kidd, as Reader of Mineralogy, needed a successor (as, also a medical doctor, he had accepted a readership in anatomy), his first choice was Conybeare, who was regarded as the brightest member of the geological circle of his day. However, Conybeare wanted to take a wife, which was forbidden to Oxford dons at the time; moreover, his inheritance gave him a certain independence in the form of £3,500 a year. Kidd's second choice, Buckland, who was less impatient to marry (he would wait another fifteen years) was therefore given the job. Conybeare, forsaking what might have been a distinguished academic career, was left free to devote himself to his passion for ancient rocks and fossils. Yet Buckland so appreciated Conybeare's originality that he sought his advice in preparing his inaugural lecture, and his help is in evidence in the eloquent and highly regarded '*Vindiciae geologicae*' that emerged.

In 1816, Conybeare, Buckland and George Greenough, first president of the Geological Society, visited Germany to study German mineralogy and geology. They met, among others, Goethe and Werner (then in his sixties), and compared the rocks with what they knew from Britain, and saw famous fossil sites such as the beds of bears' bones in a cave at Franconia in Germany. Conybeare incorporated many of their observations into his authoritative book, co-authored by his publisher, William Phillips, *Outlines of the Geology of England and Wales*. This was an important work, summarising the state of geological knowledge in his country, starting with the most recent at the top and descending to the Old Red Sandstone and, for the first time, naming the Carboniferous, which owes its name to the coal beds that were laid down globally during this period, 304 to 312 million years ago. It listed all the fossils found in each formation. The same year of their tour, Buckland published the first table comparing the strata of England with those of the Continent. Well before the Great Devonian Controversy, he called attention to the resemblance of the greywacke slates he saw to the Transition formations found on the English and Welsh borders.

Buckland preached geology as a kind of religion, in which its hammer blows were forcing rocks to give up their secrets, or, as he put it in his inaugural lecture: 'it is surely gratifying to behold Science, compelling the primeval mountains of the Globe to unfold the hidden records of their origin'.[11] He lectured on geology from 1814 at Oxford and had the distinction of being the first to teach a geology course at an English university (though the subject had been taught in Edinburgh since 1781). In 1818 his grasp of the subject was recognised by the Royal Society, which made him a fellow. He also became director of Oxford's Ashmolean Museum of Natural History.

Buckland would become a popular and humorous lecturer, waving bones and fossils about to illustrate his argument. University lectures had to be entertaining or nobody would attend them. Natural science, as Charles Lyell had found at King's College London, was extracurricular – that is, optional. There were no examinations. Buckland had to work hard to persuade students to come to hear him and he badly needed the money from the fees they paid for attendance.

Sometimes Buckland would draw an audience by lecturing outdoors in a quarry or on a hill to demonstrate stratigraphy in situ, after which he would gallop off, sometimes sporting a fossil around his neck. He kept beside him his large blue sack of fossils and accoutrements from which he was never parted, even at formal dinners. And his stunts were legendary. Behind a large showcase he would pace up and down, holding, say, a huge hyena's skull. According to one student's account, he suddenly dashed down the steps, skull in hand and shouted at a young man on the front bench: 'What rules the world?' The youth shrank back. Buckland rushed at another, pointing the skull straight in the student's face, and got the reply: 'Haven't an idea.' Buckland, mounting his rostrum, announced triumphantly: 'The stomach, sir, rules the world. The great ones eat the less, and the less the lesser still.'[12]

Conybeare scoffed at Buckland's fondness for making a religion of geology. In a sarcastic poem, 'Ode to a Professor's Hammer', he wrote:

Hail to the hammer of science profound! . . .

Beneath the storm of its thundering blows
Bending, and opening, and staggering, and reeling,
Mountains reluctant their story disclose,
The secret of millions of ages revealing.

The fossil dead that so long have slept,
And seen world after world into ruin swept
Start at the sound
Of its fearful rebound.[13]

Buckland never disappointed his audience. In Oxford in 1832, at the second meeting of the British Association for the Advancement of Science, he gave a theatrical performance on the subjects of geology and the courtship of primitive reptiles. (The first meeting of the association that became familiarly known as 'the British Ass' had been held the previous October, 1831, at York where the subjects discussed were comets, railways, geological strata, the Aurora Borealis and marsupial mating habits.[14]) At the Oxford meeting Buckland used the occasion for one of his jokes. Behind him, as he spoke, stood a skeleton of a giant fossil, a megatherium, from Argentina. He invited the geologist William Clift to step through the bones and 'come a second time into the world through this cavity in the pelvis of the megatherium'.[15] There was great applause and the session went on until midnight. *The Times*, however, dismissing the lecture as 'a mere unexplained display of philosophical toys', wrote reprovingly that Buckland sometimes seemed to forget that he lectured in the presence of ladies.[16]

He was also a formidable party-giver. Drink flowed freely at Buckland's gatherings and, as always, he worked hard to amuse his guests. A fossil skeleton of an *Ichthyosaurus* served as a candelabrum. The jumble of bones, stones and animal relics in his rooms in Corpus Christi was celebrated. A wit mocked them in a poem, 'Picture of the Comforts of a Professor's Rooms in C.C.C., Oxford':

> Here see the wrecks of beasts and fishes,
> With broken saucers, cups, and dishes . . .
> Skins wanting bones, bones wanting skins,
> And various blocks to break your shins.[17]

His boast of eating his way through the animal kingdom was celebrated. No form of meat or fish was too bizarre to stop him tasting it. Hedgehog and crocodile were among the delicacies he fed his guests. The poet and artist John Ruskin, who as an Oxford undergraduate had been invited to Buckland's table, wrote in his journal: 'I have always regretted a day of unlucky engagement on which I missed a delicate toaste of mice.'[18] Later, when Buckland was living in London as Dean of Westminster (a post he accepted in 1845, disillusioned with Oxford's attitude towards science), his dinners at the deanery featured horse's tongue, alligator, puppies, mice, tortoise, bison and ostrich. He even claimed to have eaten the heart of King Louis XVI, sold by a scavenger in Paris to his friend Lord Harcourt at Nuneham. According to legend, he devoured the whole organ, leaving nothing for its collector.

In 1822 Buckland achieved fame. He owed it to his exploration of a cave in Yorkshire where he found a rich haul of remains of extinct hyenas.

The cave had been discovered accidentally, when the ground gave way beneath a workman and swallowed his pick. When Buckland got to the cavern, he saw what he discerned as the bones of hyenas, wolves and bears. Only the hyena bones were intact, suggesting to him that the hyenas had killed the other creatures and dragged the remains back to their den to gnaw the bones. Previously such bone heaps had been taken as evidence that the Noachian Flood had swept tropical animals to northern climes. But that explanation did not explain why the bones had been chewed.

Exploring Kirkdale Cave was no easy exercise. Holding a lantern, Buckland had to creep on all fours into the dark sloping hole two feet wide and five feet high. Then he had to slide over a foot-thick layer of bone-filled mud before entering dark passages that stretched hundreds of feet back into the hillside.

The effort was rewarded. He came across the fossil bones of disparate animals including extinct species of hippopotamus and elephant and the skull of a rhinoceros; all appeared to have been broken. By the Great Flood? But there was no sign of water damage. Had they been gnawed? If so, by which animals? Buckland found his explanation in more than three hundred hyena teeth and the bones of seventy-five hyenas. He reasoned that the larger animals could not have been washed into the cave by the Flood as the opening was too small; they must first have been killed, then carried in parts by the hyenas, which then gnawed them; when the waters swept in, the entrance was sealed off.

Buckland also found coprolites. He was fascinated by these small dark-grey round balls of calcified excrement, which appeared to have been preserved far more than the bones of the animals that had produced them. Despite the prevailing prudery of his time, he showed a complete lack of embarrassment when discussing them. He identified the droppings found in the Kirkland Cave as *album graecum* – white deposits that were the petrified faeces of extinct vertebrates. To him, they were just as fascinating as the bones and stones. When the keeper of a menagerie which included hyenas supported Buckland's hypothesis, Buckland responded delicately that 'though such matters may be instructive and therefore to a certain degree interesting, it may be as well for you and me not to have the reputation of too frequently and too minutely examining faecal products'.[19]

Buckland's well-known interest in ancient excrement inspired De la Beche's witty etching entitled *A Coprolitic Vision*. It shows the 'Reverend Professor of Mineralogy and Geology in the University of Oxford', wearing gown and mortar board, with a geological hammer in his right hand, standing on a flat rock at the opening of a wide cavern as if it were an auditorium where he is conducting a performance by a deer, a bear, a leopard, hyenas, crocodiles, ichthyosaurs and pterodactyls. Each creature has lumps dropping from its backside, some landing near the reverend professor's feet.

Buckland's paper on the Kirkdale Cave received a rapturous reception when he read it at the Royal Society in 1822 and won him the society's highest honour, the Copley Medal, never before given to a geologist. In his welcoming speech, Sir Humphry Davy praised the paper as marking the moment when geology caught up with astronomy. The new science, he said, had reached the point 'from which our researches may be pursued through the immensity of age, and the records of animated nature, as it were, carried back to the time of the creation'.[20] Buckland was delighted by the recognition. He wrote to his friend Lady Mary Cole: 'The president and Council of the R.S. have sanctioned my paper with the Copley Gold Medal so that I am now not much afraid of any further opposition to my Hyena Story which my friends at first predicted no body would believe . . .'[21]

The Kirkland Cave paper inspired another famous cartoon, this one attributed to Conybeare. It shows Buckland poking his head into a prehistoric hyena den, holding a candle while the live hyenas look up from their gnawed bones and menace the intruder. It was captioned: 'Mr. BUCKLAND peeping into the Hyaenas' den, to see what they are about, etc.'. Conybeare was unquestionably the author of the accompanying poem with the unpoetic title, 'The Hyaenas' Den at Kirkdale near Kirby Moorside in Yorkshire, discovered AD 1821':

Their teeth had the temper of steel,
Skulls & dry bones they swallowed with zest, or

Mammoth tusks they dispatch'd at a meal,
And their guts were like Pappin's digester.

Buckland enjoyed the poem and the drawing, which portrayed him as a spy creeping into the past – a time-traveller, with hair standing on end. The poem, printed as a broadsheet accompanied by the drawing, was leaked to the *Yorkshire Gazette*.[22] The sketch has been held as an even earlier candidate than De la Beche's *Duria Antiquior* as the first artistic attempt, based on new geological knowledge, to recreate the reality of the ancient past – not through fictional dragons but through representations of extinct animals as they had appeared when alive.

There was nothing jokey, however, about Buckland's use of the Kirkdale evidence in his Oxford lectures. A lithograph of him lecturing to senior members of the university shows him holding a fossil hyena jaw over a desk full of fossils with Greenough's geological map of England in the background. In it he is clearly an academic lecturer of utmost seriousness.

As an ordained priest and theologian Buckland was well aware that 'dating the deluge' was a major preoccupation among his fellow clerical geologists. He himself believed that there had been a huge watery catastrophe so recent that it could be interpreted as the biblical Flood. As he saw it, an extraordinary wave of water had swept across Europe, making extinct the hyenas and other animals found in the Kirkdale Cave. In 1823 he put these arguments into what became his *Reliquiae Diluvianae*.

Before the book was done, however, he found an even better cave for his purposes: Paviland Cave in cliffs facing the sea on the south coast of the Gower Peninsula, west of Swansea. The cave, known locally as 'the Goat's Hole', was (and is) accessible only at low tide.

From his friend Lewis Weston Dillwyn, a Swansea businessman, naturalist and fellow of the Royal Society, Buckland had learned that a trove of bones had been found inside Paviland Cave. On 24 December 1822, Buckland contacted Lady Mary Cole, the remarried widow of T. M. Talbot, former head of the wealthy Talbot family who lived nearby at Penrice Castle. (She was, with her daughter Mary Talbot, one of the upper-class women who often became amateur naturalists at that time.) He begged her to send him 'a few of the best marked teeth & Bones in a Box by the Mail'.[23] She went with two male companions and collected 'a great quantity' of bones which she had sent to Buckland.

In the third week of January 1823, Buckland then went to Swansea to see for himself. He soon found that Paviland Cave would be even harder to enter than Kirkdale had been. 'Visitors reach it by scrambling up the rocks, which are uncovered for two and a half hours either side of low water,' Paul Ferris explains in his history of the Gower. 'Anyone caught by the incoming tide is safe enough there, but has a seven-hour wait before they can get out again. The only other route is the face of the cliff, which is dangerous.'[24]

When Buckland, accompanied by one of Dillwyn's sons and several others, entered Paviland Cave, they unearthed, beneath six inches of mud, a human skeleton. Lying in extended burial position, its bones were identifiable as coming from the left side of the body: upper arm, inner forearm and bone of the forearm extending from the lateral side of the elbow to the thumb side of the wrist. The pelvis lay in place as did the entire left leg and foot, part of the right foot, and many ribs. Most striking was the reddish colour of the bones. Nearby were fragments of ivory rings and curved ivory rods. They also uncovered (nothing escaped Buckland's eye) a mutton bone that seemed to have been used as a primitive instrument. The visitors assumed the skeleton to have been a murder victim.

In all, Buckland made three visits to Paviland Cave. Discussing the find with his friends, he at first joked that the skeleton might have been that of a tax man murdered by smugglers. Then his imagination got to work. The mutton bone, he decided, was a conjuring tool used by a witch, as were the ivory rods and rings. A quick mental leap led him to decide that the remains were those of a female and that their red colour indicated that she had been not only a witch but also a prostitute. Buckland amplified this theory by pointing out the cave's proximity to an Iron Age enclosure, or fort, at the top of the hill. Therefore, 'whatever may have been her occupation, the vicinity of a camp would afford a motive for residence'. He dubbed her 'the Red Lady or the Witch of Paviland' and made her a highlight of what would be his best-known work, *Reliquiae Diluvianae (Relics of the Flood): Observations on the Organic Remains attesting the Action of a Universal Deluge*.

Later scientific analysis showed that this celebrated skeleton was that of a male aged between twenty-five and thirty. The red stain on the bones has been traced to the ochre found in local rocks nearly 30,000 years old. But facts have not dimmed the Red Lady's fame. She – or rather, he – lies in the Oxford University Museum of Natural History.

In 1824 Buckland was elected to the first of his two terms as president of the Geological Society. (The second ran from 1839 to 1841.) He was president when, on 20 February 1824, the society got its royal charter, which brought, as well as high social status, legal benefits such as the right to enter into contracts and to hold bank accounts. Only five national learned societies shared this distinction: the Royal Society (in 1660), the Society of Antiquaries (in 1751), the Royal Society of Edinburgh (in 1783), the Royal Irish Academy (in 1785), and the Linnean Society (in 1802).

Perhaps it was Buckland's good sense that kept the Geological Society from appending 'royal' to its name and annoying the Royal Society. In any event, he hosted a celebration dinner on the winning of the charter, appropriately at the Freemasons' Tavern where the society had been formed.

On that occasion, Buckland had good reason to celebrate, for at the meeting – the same at which Conybeare presented his description of the plesiosaur – Buckland announced the discovery of a giant reptile with the serrated teeth of a carnivore and the legs, sacrum and vertebral column of a mammal. The find was all the more remarkable because it included a part of a lower jaw with curved and pointed teeth still in place. The gigantic bones had actually been found years earlier, in a stone quarry surrounding the village of Stonesfield, eight miles northwest of Oxford, long known for its richness in fossils. The bones had been brought to the Ashmolean Museum and Buckland, when he became director, by careful studies, had made them his own. He told his audience the name, meaning giant lizard: 'I have ventured, on concurrence with my friend and fellow labourer, the Reverend W. Conybeare, to assign to it the name *Megalosaurus*.'[25] It was, although none realised it at the time, the first scientific description of a dinosaur. It was also one of the first proofs that in the remote past giant reptiles had roamed the earth as well as inhabiting the sea. (In 1842 the comparative anatomist and palaeontologist Richard Owen assigned *Megalosaurus* and other ancient beasts to his new order, Dinosauria – terrible lizard.)

The giant bones had already brought the famed Cuvier over from Paris in 1818 to see what was at first termed *Megalosaurus bucklandii*. Cuvier, who had seen similar large bones in Normandy, returned to Paris all the more convinced that the earth had once been inhabited by giant reptiles. He set about expanding his classic *Ossemens fossiles*, with illustrations of a great array of specimens. He gave particular attention to the fossils of large beasts such as the rhinoceros and hippopotamus as well as mammoths and mastodons, and in his final volume revised and expanded his earlier work on fossil reptiles. Geology, as he now saw it, was a historical science.

What remained to be discovered was how and when the age of reptiles preceded the age of mammals.

A year after his '*Megalosaurus* . . . of Stonesfield' paper, influential friends arranged for Buckland, aged forty-one, to become a canon of Christ Church Cathedral. This promotion allowed him to move out of the college and into a good and free house, with an income of £1,000 a year. It also allowed him to 'enter into the holy estate'. He wasted no time. In 1825 he married Mary Morland from Oxfordshire, a fossil geologist who had a great talent for drawing and sketching.

Buckland had first spotted Mary while travelling by coach to Dorset; he was reading a heavy tome by Cuvier and saw a woman passenger reading the exact same book. On speaking to her he was surprised to find that she had worked as an illustrator on the book. By chance, on that journey, Buckland carried with him a letter of introduction to one Mary Morland as someone he might like to meet in Dorset. They were married by the end of the year and spent a geological honeymoon inspecting caves in France and Sicily, before settling down at Oxford. They would have nine children together.

Buckland now concentrated on dating the deluge. In 1836 he contributed to the *Bridgewater Treatises*, an important series designed to help the general public understand science as a manifestation of the wisdom of God. His map, which he incorporated, contained a diagram of an 'ideal section' of the earth's crust and shows that he accorded a vast age to the rocks that had preceded human life. In his book, *Geology and Mineralogy, Considered with Reference to Natural Theology* (1836), he pushed back the last geological debacle from the biblical deluge to the period immediately preceding the creation of man. He declared the language of rocks and fossils to be as much a divine revelation of truth as the Bible. As he presented it, geology made up a history of a high and ancient order, written by the finger of God himself upon the foundations of the everlasting hills. This book was so popular that its first edition print-run of 5,000 (far greater than the 1,500 printed for the first edition of Charles Lyell's *Principles*) sold out before it reached the bookshops, despite being the same hefty price of thirty-five shillings. It went on to sell thousands of copies more.

In time, however, Buckland's reputation declined. His former pupil, Lyell, was unconvinced by Buckland's assertion that geology testified to God's design and that fossils illustrated the progress of life as it developed towards its divine destination: man.

Lyell consistently and sternly resisted any attempt by Buckland and other clerical geologists to explain the Noachian deluge in geological terms. Using the editorial 'we', Lyell wrote: 'For our own part, we have always considered the flood, if we are required to admit its universality in the strictest sense of the term, as a preternatural event far beyond the reach of philosophical inquiry.'[26] Those intent on equating geological phenomena with biblical history, he advised, should remember the remarkable fact that in the Bible a dove flies back to the Ark bearing an olive branch (Genesis viii:11). This occurrence could only mean that the earth had not been universally submerged and that somewhere on the earth, vegetation had survived.

The Flood, however, was still a delicate issue (and so it remains today for the many who insist on biblical explanations for the earth's history). Lyell, frowning on 'diluvial interpretations',[27] argued that the biblical flood was taken to have been a universal violent rush of waters and to have covered the whole earth, including 'the summits of the loftiest mountains', and that the multitude of volcanic cones in central France must have arisen since the date usually assigned to the deluge (about 4,000 years BC). By that timing, the mountains around Auvergne, with their well-preserved craters, had to have appeared since the beginning of the recorded history of France – perhaps even since the conquest of Gaul by Julius Caesar in 51 BC; it was impossible to believe that

Caesar could have observed these dramatic eruptions and hot lava flows without mentioning them.

Lyell hoped that his work would not offend orthodox Christian thinkers and 'would only offend the ultras'.[28] He succeeded. The religious periodical press welcomed *Principles* and called for its study by all those interested 'in the Great First Cause'.[29]

The effect of Lyell's book on the Geological Society was profound. It persuaded them to renounce the equation of the Flood with the many traces of fluvial erosion, deep valleys and watery catastrophe found around the earth.

Should Buckland be re-evaluated? He was not universally admired in his time. Charles Darwin wrote in his autobiography that he knew and liked all the leading geologists, 'with the exception of Buckland, who though very good humoured and good natured seemed to me a vulgar and almost coarse man. He was incited more by a craving for notoriety, which sometimes made him act like a buffoon, than by a love of science.' Buckland's reputation suffered also from the mental illness that clouded his last years and had him sent to an asylum.

Yet it has been argued that more than anyone else, Buckland made 'deep time' acceptable to contemporary thinkers and opened the way to ideas about evolution. In the *Times Literary Supplement* in 2008, the eminent geological historian Richard Fortey wrote persuasively that William Buckland has been 'too long tarred with the error of eagerly recognising evidence of the Mosaic Deluge in the fossil bones he dug up from limestone caves. But he soon left the theory behind as new evidence came to light, and he was a crucial figure in showing how the living detail of the prehistoric past could be revived through critical study of fossil bones. He deserves to be recognised as one of the pioneers in re-animating the ancient earth. He discovered the oldest human fossil. One would like to have met him.'[30]

7

ON THE BEACH

The Cobb, a long grey wall curving out to sea, is the outstanding feature of Lyme Regis in Dorset. John Fowles's novel *The French Lieutenant's Woman* calls it 'quite simply the most beautiful sea-rampart on the south coast of England . . . a last bulwark against all that wild eroding coast to the west'.[1]

The coast of Lyme Regis was the birthplace and inspiration of Mary Anning. Her discoveries in the first half of the nineteenth century have led to her being acclaimed in her lifetime as the greatest fossil-hunter the world has ever known.

Born in 1799, from youth to middle age Anning was a familiar figure on the Lyme shoreline in her long skirt, shawl, bonnet and basket, endlessly toiling to find the treasures she knew were buried in the unstable rocks and in the sands uncovered at low tide. With phenomenal skill and knowledge, she combed the cliffs and the beach for marine fossils such as ammonites (spiral-shaped rocks once called 'snake stones') and belemnites (small, bullet-shaped invertebrate fossils), believed to have magical or curative properties. Anning was adept not only at spotting and extracting the fragments but also at cleaning and polishing them, so that their details were best revealed. Her speciality came to be known as palaeontology.

Today Anning stands high on the list of once-forgotten, now-revered females. A familiar feminist tale can be made of her life: she did the work, the men got the credit. But that is a caricature. Low social class and poverty far more than gender are what kept Mary Anning out of the London geological community, dominated by wealthy Anglican gentlemen. Yet in the twenty-first century she has become an icon – the subject of biographies, a novel, and a costumed re-enactment at London's Natural History Museum.

Her family, who lived in a house on the town's bridge, were religious Dissenters. She learned to read at the Sunday school run by the Congregational Church and before long was able to study scientific papers and perform dissections to understand the anatomy of the fossils on her table. From an early age, she knew that any heavy rainstorm was likely to expose bones buried in the chalk. Her first big find came in 1811, when she was twelve.

Richard Anning, her father, was a skilled cabinet-maker. He moved to Lyme Regis in 1793 after his marriage and found himself becoming less interested in the containers than in the things contained. The strangely shaped curios found on the beach were ideal trophies to put into the display cabinets that stood in fashionable drawing rooms. Lyme Regis itself was a popular tourist resort booming with the new activity of 'sea-bathing', performed for health rather than for exercise. (Bathers' modesty was protected by bathing machines that allowed them to splash in the salt water, uncovered bodies unseen.) Richard Anning, with a ready market provided by tourists seeking souvenirs and refuge from the rain, set up a stall in front of his shop and filled it with sea treasures. He taught fossil-hunting to his son and daughter, who accompanied him as he searched the cliffs.

When Mary was ten and a woman gave her half a crown for an ammonite, she was on her way. She learned early the practicality of strapping pattens – a protective overshoe – over her ordinary shoes to keep her feet raised above the damp sand. Indeed, long skirts were impractical too, but in the early nineteenth century, trousers were not an option for an English female.

The Anning family had a troubled life. Seven of their children died early, the eldest in a fire at the hearth in their home. Richard Anning died of consumption in 1810 at the age of forty-four, having been weakened by a fall from a cliff. A baby born the following year died at birth. Thus the family was reduced to three: Mary, her older brother, Joseph, and their mother, Mary Moore, for whom her surviving son and daughter were the only hope of staying out of the poorhouse.

Fossils came to the rescue. To catch the eye of tourists, Mary and Joseph set up a table of stony curios (they called them 'curiosities') near the coach stop at a local inn. As the growing enthusiasm for geology had raised the price of fossils, the pair were more successful than their father could have dreamed.

In 1811 Joseph uncovered a huge skull protruding from a cliff. It was four feet long, with gaping eyes and more than 200 teeth. Because Joseph was committed to an apprenticeship to an upholsterer, he asked his sister to try to find the rest of the skeleton. She tried hard but as that part of the beach became covered by a mudslide, she ended up working on the project for nearly a year. (The whole town knew of it, especially because of her reputation as a girl 'blessed by divine favour': as a fifteen-month-old infant she had survived a bolt of lightning that killed the young neighbour holding her and two other girls as well – a stroke of fortune that made her something of a legend in Lyme Regis.) Digging away, with some help from local quarrymen, Mary uncovered virtually the entire connected skeleton whose head her brother had found. After local men removed it from the hillside, she cleaned it and reassembled the pieces of what turned out to be a giant reptile, with four fin-like flippers, a large tail and – perhaps its most distinctive feature – a ring of bony plates surrounding a large eye. Nothing like it had been seen before.

Assembled, the skeleton with the gaping eye sockets stretched to seventeen feet. The lord of the manor, Henry Hoste Henley, bought it for £23, then sold it to the London collector William Bullock for his Museum of Natural Curiosities at 22 Piccadilly. There it was placed in the new 'Egyptian Hall' where crowds flocked to stare at it. The startling fossil was described for the Royal Society in six scientific papers (none of which mentioned either Joseph or Mary

Anning's name) and it was assigned to a new reptilian genus called *Ichthyosaurus* – fish lizard. In 1819 the British Museum bought it for £24 and accepted the *Ichthyosaurus* name.

This spectacular fossil, revealed in the same early decade of the nineteenth century that produced William Smith's great map, was yet one more new inescapable statement of the great age of the earth.

In 1820 a local patron, Lieutenant Colonel Thomas Birch, organised an auction of specimens he had bought from the young Annings. The sale attracted much publicity and raised £400 for the Anning family.

As a teenager Mary made the acquaintance of the first of the important geologists who would admire her achievements. Henry De la Beche was then a young man, years away from the Great Devonian Controversy that later engulfed him; he was then merely a handsome sixteen-year-old who had been expelled from a military college for insubordination and who had moved to Lyme Regis with his newly remarried mother. He and Mary combed the cliffs together. Later he publicised her discovery of the new genus, *Ichthyosaurus*, at the Geological Society.

It was sometime around 1815 that Mary Anning first met William Buckland. The then unmarried Oxford professor often came down to Dorset on his black mare, carrying his blue bag and hammers, to explore the coast he knew so well from childhood. On one visit he looked up the now-famous young fossil-hunter. There was talk in Lyme Regis about the pair being seen often on the beach together. Years later, Buckland's daughter recalled that 'local gossip preserved traditions of his adventures with that geological celebrity, Mary Anning, in whose company he was seen wading up to his knees in search of fossils in the Blue Lias'.[2]

Mary Anning went on to find new and more complete skeletons of ichthyosaurs. In 1823 she uncovered the complete skeleton of the long-necked plesiosaurus – the giant specimen that Conybeare was unable to get up the stairs at the Geological Society a few months later. 'Find' is the wrong verb. It had taken her ten years to dig it out. The Duke of Buckingham, having first checked its authenticity with Conybeare, had agreed to pay Mary Anning £110 for the skeleton – then the highest price ever paid for a fossil.

Her discoveries escalated. In 1828 Anning's work was noticed in the *Bristol Mirror*. The new Bristol Institute had bought an ichthyosaur skeleton she had found. The article recounted the difficulties of her work: 'This persevering female has for years gone daily in search of fossil remains of importance at every tide, for many miles under the hanging cliffs at Lyme, whose fallen masses are her immediate object, as they alone contain these valuable relics of a former world, which must be snatched at the moment of their fall, at the continual risk of being crushed by the returning tide – to her exertions we owe nearly all the fine specimens of *ichthyosauri* of the great collections.'[3] The risk of her work was indeed considerable. There was the constant danger of landslides. One fell on her beloved dog, Tray, burying him, and despite her frantic efforts to dig him out, he did not survive.

In 1829 she uncovered a squaloraja (a fossil fish, transitional between a ray and a shark). In the winter of 1830 she made her fifth major discovery: another new species of plesiosaur, which came to be called *Plesiosaurus macrocephalus* for its extraordinarily large head.[4] She wrote to Buckland and De la Beche that 'it is without exception the most beautiful fossil I have ever seen'.[5] (It was bought by a Conservative MP for £200.) She never made any attempt to publish her findings.

Anning kept up her pace of discovery. In 1828 she found a pterosaur, the first flying reptile ever found in Britain. In 1829 she found a winged skeleton with a large skull, a long beak, two types of teeth – long in the front, shorter sharp ones at the back, and four fingers: in other words, a gigantic flying lizard. Buckland travelled down to Lyme Regis to see it. He then broke the news for the Geological Society and at last mentioned her name: 'Miss Mary Anning . . . has recently found the skeleton of an unknown species of that most rare and curious of all reptiles' – he meant a flying lizard, a pterosaur.

Buckland did not stop there. Rapturously he drew comparisons with a crocodile, a lizard, a bat, an iguana: 'in short, a monster resembling nothing that has ever been seen or heard-of upon earth, excepting the dragons of romance and heraldry'.[6] Today it is identified as *Dimorphodon* – a creature with two kinds of teeth.

Anning's discoveries became incontestable evidence for the extinction of species. Until then most geologists, such as Lyell, believed that animals did not become extinct but might be found somewhere else on earth. The bizarre nature of Anning's fossils opened the door to understanding life as it had existed in the distant ages of geological history. At a time when London Zoo, opened in April 1828, gave visitors their first sight of elephants, giraffes and rhinoceroses, the public was prepared to be told that the earth was once inhabited by gigantic reptiles.

Anning's reputation spread abroad; she had many visitors who came to see the famed fossilist as well as the famed coast. In 1844 King Frederick Augustus II of Saxony visited her and paid her £15 for a fossil of an ichthyosaurus six feet long. Supplying her name and address to the Saxon king's personal physician and guide, she wrote beside it: 'I am well known throughout the whole of Europe.'[7]

She still needed money though. Her health was poor: in her late thirties, in recognition of her research and predicament, Anning was granted an annuity raised jointly by the British Association for the Advancement of Science and the Geological Society. Many distinguished Europeans clamoured to meet her and to see her finds.[8] In 1830 her original findings were memorialised in the classic art work of early geology. Her friend, now an artist, Henry De la Beche arrived on her doorstep and presented her with a copy of his *Duria Antiquior*, which depicted, as she would have recognised, her personal discoveries: three types of ichthyosaur, a plesiosaur and the *Dimorphodon*. He had created the piece specifically for Anning's benefit and gave her the proceeds.

As far as is known, Anning never had any suitors. Yet she was not immune to male charms. She told Charlotte Murchison, wife

Sir Charles Lyell, 1855.

Lady Mary Lyell, 1860s.

'Boulders drifted by ice on shores of the St Lawrence' from Charles Lyell's *Principles of geology; or, The modern changes of the earth and its inhabitants considered as illustrative of geology*, published by John Murray in 1875.

A 1787 print showing James Hutton hammering at a rock face made up of the faces of his opponents.

Strata of red sandstone, slightly inclined, Siccar Point, Berwickshire. From *A Manual of Elementary Geology* by Charles Lyell, 1852.

Georges Cuvier, 1822.

Sir Humphry Davy, c. 1830s. Engraving by Edward Scriven after the 1821 painting by Sir Thomas Lawrence.

Wikimedia Commons

William Conybeare's 1822 cartoon of William Buckland entering a prehistoric hyena den.

William Buckland equipped to explore a glacier, 1875.

TERTIARY SERIES AND EXTINCT VO
Land Plants
Chalk
Granite
Grauwacke
Primary Slate

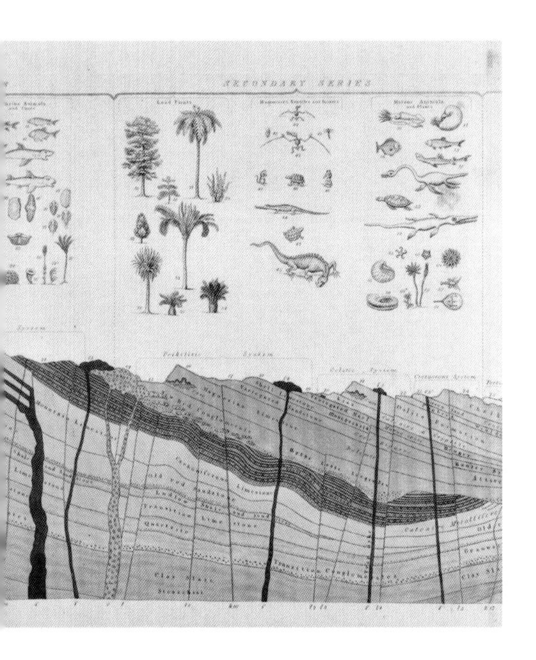

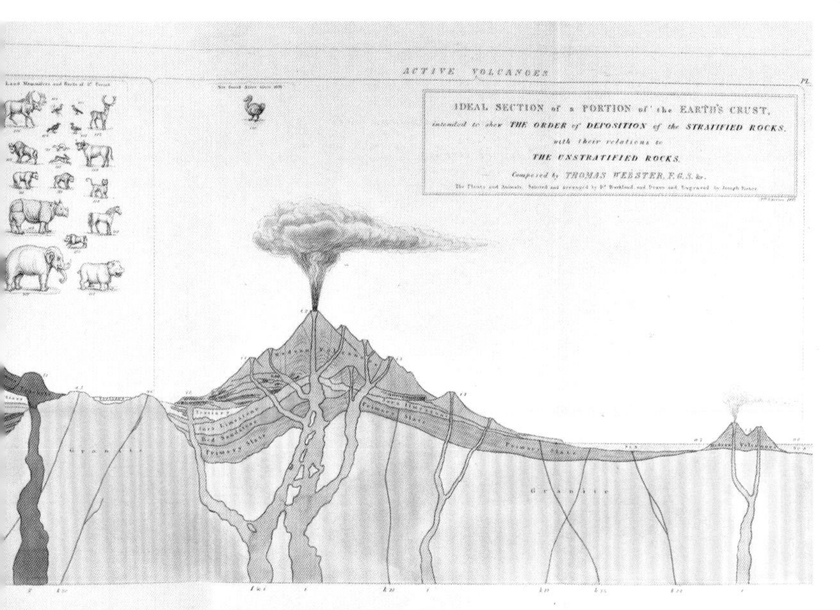

Buckland's wall map, the frontispiece to his *Bridgewater Treatise*, first published in 1836.

Sketch of an Ordinary Meeting of the Geological Society, probably by Henry De la Beche, c. 1830.

Punch cartoon from 1858, mocking the Victorian fashion for fossil-hunting.

of Roderick, that her husband was 'certainly the handsomest piece of flesh and blood I ever saw'.[9] That did not spoil their friendship. When the Murchisons first visited in 1828, Roderick told his wife to stay in Lyme for a few weeks to 'become a good practical fossilist, by working with the celebrated Mary Anning of that place'. The two women became lifelong friends and it was Charlotte, with her wealthy connections, who helped Mary reach collectors in Europe.

The following year, Anning, having seen her brother marry, visited London for the first and only time in her life. The Murchisons had long urged her to come and in 1829 she stayed with them in elegant Belgrave Square. She visited the British Museum and was bewildered by the luxury and poverty and bustle of the capital.

Her work progressed but her health continued to fail. In 1846 Buckland persuaded the Geological Society to create an additional fund for her. In that year also she was named the first honorary member of the new Dorset County Museum in Dorchester. Her illness was breast cancer, and she suffered, often bedridden, for another two years before dying at the age of forty-seven.

By the time of her death she had, like her brother Joseph, left the Dissenters and joined the Anglican church of St Michael's in Lyme Regis. Although undoubtedly not her intention, the conversion allowed her to be buried in the churchyard overlooking the sea, with a stained-glass window (to whose cost the Geological Society contributed) commemorating both 'her usefulness in furthering the science of geology' and 'her benevolence of heart, and integrity of life'.[10]

In 1848 De la Beche, at the end of his presidential address to the Geological Society (he was also director general of the new Geological Survey) recalled the 'talent and good conduct' of Mary Anning who, 'though not placed among even the easier classes of society, but one who had to earn her daily bread by her labour, yet contributed by her talents and untiring researches in no small degree to our knowledge of the great *Enalio-Saurians*, and other forms of organic life entombed in the vicinity of Lyme Regis'.[11]

Life was kinder to De la Beche himself. He was knighted in 1842 and in 1852 was awarded the Geological Society's Wollaston Medal for his efforts in turning geology from a gentleman's hobby into a professional occupation in which many found work in the British Empire.

Mary Anning, meanwhile, remains a figure of pity and awe. In *The French Lieutenant's Woman*, Fowles wrote: 'one of the meanest disgraces of British palaeontology is that though many scientists of the day gratefully used her finds to establish their own reputation, not one native type bears the name *anningii*'.[12]

Her most fitting memorial, however, is De la Beche's masterpiece, *Duria Antiquior*, depicting as if alive every creature she was the first to discover.

8

DINOSAUR WARS

It is a wonder that one of the earliest and ablest geologists found time to hammer the rocks. Gideon Algernon Mantell, a country doctor born in Lewes, Sussex in 1790, was devoted to exploring the fossil-filled strata of England's South Downs. Even so, his primary activity (to judge from his vivid personal journal) was bloodletting.

If geology was a new science when he began, medicine was an old one, reliant on techniques that make painful reading today. In 1818 James Moore, the Lewes surgeon and general practitioner to whom Mantell was apprenticed, suffered 'a most severe fit of Apoplexy'. Immediately, Mantell made 'a large orifice in a vein and took away forty ounces of blood'.[1] He then applied leeches. Three days later Moore was still having convulsions: 'I resolved to bleed him again as the dernier resort. I took from him 20 ounces; whilst the blood was flowing he muttered out "better Mantell, better": this induced me not to stop the bleeding till his pulse faltered; he amended from that very minute.'[2]

Despite the treatment, Moore recovered with no more lasting effect than a lisp from the loss of two front teeth. Mantell bought the practice from him that year. Moore, like many physicians of the time, was not only a certified obstetrician and surgeon but also an apothecary – a vocation that led him, like Mantell, to an appreciation of the medicinal properties of minerals and fossils to be found in the earth.

In 1811, with some money left to him by his father, Mantell went to London for medical training at the Royal College of Surgeons, and later qualified as a surgeon and obstetrician. While he was in

London, his interest in fossils was encouraged by an older doctor with a similar obsession. James Parkinson, the distinguished surgeon of St Bartholomew's Hospital, began acquiring fossils around 1798. Parkinson, an early member of the Geological Society, put his knowledge into a large book, *Organic Remains of a Former World*. Its very title shows the revolution in thought that fossil awareness had brought to the early nineteenth century. Thanks to Parkinson, Mantell became one of the first to appreciate the significance of William Smith's great 1815 map based on stratigraphical sequences across England.

That same year, Mantell achieved his first published work (even if only in a local newspaper). He called it 'On the Extraneous Fossils found in the Neighbourhood of Lewes'.[3] For the same newspaper he wrote a letter on vaccination which was widely noticed. Gradually he accumulated an outstanding fossil collection, in part by buying from a London dealer. He became well acquainted with the work in France of Baron Cuvier and his colleague, Alexandre Brongniart, who had found the remains of large animals in the environs of Paris. Mantell turned himself into an expert on bones – human and animal – with a fascination akin to Cuvier's for parts and their conjoining. Like De la Beche, he also had a gift for drawing.

Back in Lewes after his London training, Mantell attempted to devote himself to the project in which his heart lay – a book 'on the Geological structure of this County'. It was published in 1822 with the better title *The Fossils of the South Downs*. Yet almost every day of his life, geology was pushed aside by medical practice, particularly obstetrics; he delivered anything between 200 and 300 babies a year.

Mantell was born in the same stretch of southeastern England which inspired Conybeare, De la Beche, Buckland and Mary Anning. He appreciated the importance of Anning's work, especially the ichthyosaur she found in 1811. His own first discovery, when he was a schoolboy, was an ammonite. He saw the rocky spiral-shaped fossil as a 'petrified serpent'.[4] The ammonite was the beginning of what became the admired 'Mantellian collection'.

Mantell was a tall, pleasant-looking, slender-faced man, whose dark hair suggested the family's Norman origin many centuries back. In his hometown of Lewes, the name was pronounced with the accent on the first syllable: 'Mantle', as if it were a cloak. Gideon's father was a shoemaker from a working-class family of Dissenters. Thus as a boy Gideon could not attend Church of England schools and had no hope of university. From the start he knew he had to find a trade. The solution was to become apprentice to the Lewes surgeon James Moore.

He had formally entered the London geological world in 1813 when he was elected a fellow of the Linnean Society of London, where he had read a paper on a fossil found in the Upper Chalk near Lewes. He consolidated his position in 1814 when he met George Bellas Greenough, the founding president of the Geological Society. In 1816, aged twenty-six, at St Marylebone Church, Mantell wed the young and beautiful Mary Ann Woodhouse, the daughter of one of his first patients. He took her back with him to Lewes for what would turn out to be a stormy marriage.

Shortly after the birth of their first son, in March 1820, Mantell learned that two years earlier he had been elected a fellow of the Geological Society, but somehow had never been notified of the honour.[5] Charles Lyell also entered his life at that time. Rock-hunting in west Sussex in 1821, Lyell was directed by some labourers to 'a monstrous clever man in Lewes, a doctor' who knew all about fossils and who dug curios out of the cliffs with which to make medicine, or, as they called it, 'physic'.[6] Mantell was delighted to meet the fine-looking young Scot and the two men sat up until the small hours talking geology. Mantell was impressed by Lyell's knowledge of rock formations in France and Italy. (They may have discussed Lyell's conviction that Britain had once been connected to France by a land bridge which, eroded over time by in-flowing currents of the North Sea, gave way to the Straits of Dover[7] – a view derived from Werner and Buckland, among others.)

Mantell's *Fossils of the South Downs* announced something that had not been widely realised before: that the beds from the area known as 'The Weald' (Old English for 'forest'), covering roughly the chalk escarpments between the North and South Downs, were of fresh-water origin. Lyell so admired the book that he mentioned it in each edition of his *Principles of Geology*.

Tilgate Forest near Cuckfield, a west Sussex village on the coach route from Brighton to London, was Mantell's favoured locality for exploration. It was there in 1822 that he made the most significant discovery of his life. According to a story familiar in the history of geology, it was his wife who made it. Mary Mantell, who illustrated *Fossils of the South Downs*, was waiting for her husband to return from a medical call when she saw on the roadside a giant tooth embedded in a piece of rock. She was well aware that her husband had been collecting such relics and bones, with the anatomist's intention of reconstructing them to their original form. Mantell himself went on to find more big teeth at the site, their size indicating that they had come from an extraordinarily large creature. He traced the source to a quarry at Cuckfield which lay in Cretaceous rock, younger than the Jurassic strata in which the Oxford geologist William Buckland's giant Stonesfield bones had been found.

The size and the fluted, flattened shape of the teeth told Mantell, a bone expert, that they must come from a land animal. However, none before was known that could chew its food. So Mantell asked Lyell to take one of the teeth with him to Paris to show Cuvier. Cuvier had no clear analysis for the tooth but suggested that perhaps it had come from a rhinoceros, but also that its source might be a different kind of animal, a herbivorous reptile. The clear answer finally came from a young assistant at the Hunterian Museum of the Royal College of Surgeons in London,* who saw in the tooth

*The Hunterian Museum in Lincoln's Inn Fields held – and holds – a vast collection of skeletons and pickled specimens, assembled by the eminent surgeon John Hunter before his death in 1793. Purchased by the British government in 1799, it was later given to the Royal College of Surgeons. The collection still attracts visitors today.

a striking resemblance to that of an iguana of which he had just assembled the skeleton. Later, comparing more of the fluted teeth, Mantell suggested to Cuvier that the creature was an ancient version of the modern iguana, a gigantic reptile (bigger than Buckland's *Megalosaurus*) and, unusually, a herbivore. The teeth were the distinguishing characteristic.

Proposing the name *Iguanosaurus*, he announced it in a paper read before the Royal Society in February 1825. Conybeare commented that his discovery was very interesting, 'but the name you propose *Iguanosaurus* will not do'.[8] Mantell accepted Conybeare's alternative: *Iguanodon* – 'don' signifying teeth. The name stuck. It was the second dinosaur to be named – Buckland's *Megalosaurus* having been the first – and the first herbivore. All other large surviving reptiles, such as crocodiles and anacondas, were carnivores – a fact that made Mantell's discovery all the more important. By 1829 he knew the importance of his findings.

It would be many years, however, before the word 'dinosaur' took hold. Richard Owen, to whom the term would be credited, was Mantell's arch-opponent. Owen was then on the staff of the Hunterian Collections of the Royal Society of Surgeons and considered himself an expert on anatomy and bones. Not until 1842, in an article for the annual report of the British Association for the Advancement of Science, which was meeting in Manchester, did Owen establish the order *Dinosauria*, relying heavily but without acknowledgement on Mantell's work. The order was intended to cover the *Iguanodon* and the *Megalosaurus*, Owen concluding that they walked on their hind legs, with two appendages like small arms higher up.[9] In consequence, Buckland dubbed Owen 'the British Cuvier' – a designation that ought to have gone to Gideon Mantell.

In 1827 Mantell published *Illustrations of the Geology of Sussex*, which is credited with being the earliest book to deal with dinosaur remains, including his own discovery. He moved to more socially prominent Brighton, where King George IV visited every winter, and forsaking medicine for geology, he established a respected geological museum. The king bought four copies of Mantell's *Geology of Sussex*. In his book Mantell stressed the enormous size of these extinct creatures: 'The gigantic *Megalosaurus*, and yet more gigantic *Iguanodon*, to whom the groves of palms and arborescent ferns would be mere beds of reeds, must have been of such prodigious magnitude, that the existing animal creation presents us with no objects of comparison.'[10]

By 1829 Mantell knew he had made his name with the *Iguanodon*. Three years later he made another major discovery, again in Tilgate quarry. It was another reptile but one composed of bony plates and spines now recognised as a form of armour. This reptile was clearly of a distinctive kind; Mantell named it *Hylaeosaurus*, meaning 'woodland reptile', and announced his discovery in a paper for the Geological Society.[11] Other dinosaur finds for which he is credited are *Cetiosaurus, Regnosaurus*, and *Pelorosaurus*. By now he had a public reputation, not only for his discoveries but for his collection held in his Brighton museum, to which he gave the grand title of 'Sussex Scientific Museum and Mantellian Museum'. Audiences of hundreds thronged to hear him expound on the life of the prehistoric past.

Mantell began a correspondence in 1834 with a professor at Yale University, Benjamin Silliman, who had been teaching geology and mineralogy as part of his chemistry course and who had travelled to Britain to study. The consequence was, as Mantell told his journal, he received: 'A letter from my excellent friend Professor Silliman, announcing that the College (Yale, Newhaven, Connecticut), has conferred upon me the degree of L.L.D. – now as Lord Byron said to Moore, "this is a way of immortality".'[12] Even so, he did not meet Silliman in person until that year.

In 1838, forced by financial necessity, Mantell sold his Brighton fossil collection to the British Museum for the great sum of £4,000 (more than £300,000 today). According to *The Times* that December, the loss of the fossils was the cause of 'great regret to the citizens of Brighton, to which they have been a most intellectual ornament'.[13] With the money Mantell bought a medical practice at Clapham in London. His wife left him, taking their four children with her. One of the two daughters subsequently died; one of his two sons, Walter, emigrated to New Zealand (where, in time, he preserved his father's papers and journal). His misfortune worsened when his spine was seriously injured in an accident in 1841 when he became entangled in the reins of a carriage carrying him across Clapham Common and he was dragged at speed over rough ground. His back never recovered and he was henceforth often in severe pain.

Mantell's later years were soured by the antagonism of the palaeontologist Richard Owen. Owen, a younger man, seemingly jealous of Mantell's reputation, seems to have lost no opportunity to disparage Mantell, even claiming for himself the discovery of the *Iguanodon*. In 1845 at the Geological Society, Owen read a paper, contradicting what he had already agreed with Mantell, saying instead that a fossil bone they had discussed was not one of a bird but of a pterodactyl. For geologists, these distinctions held tantamount to theological importance. Mantell was furious at this blatant alteration of the agreed facts. As he raged in his journal: 'It is deeply to be deplored that this eminent and highly gifted man, can never act with candour or liberality.'

Mantell's anger was heightened when the great William Buckland 'as a matter of course got up and entirely agreed in all Professor Owen stated'.[14] Mantell's own private conclusion was that the question had to be considered open until more reliable data had been obtained.

Three years later, when he read a paper on belemnites at the Royal Society, Owen once more got to his feet and 'made a most virulent attack upon me, ridiculing the communication, and stating that it was only fit for a few lines in the "Annals of Natural History"'.[15]

Even so, Mantell's paper on the *Iguanodon*, published in the *Philosophical Transactions of the Royal Society* in 1848, summarising his discoveries on that fossil reptile, won him the Royal Society's Royal Medal in 1849. It was not the society's highest honour – that was the Copley Medal – but even so the Royal Medal, sometimes known as the Queen's Medal, carried great prestige.

The envious Owen was still up to his tricks in 1850 when he dismayed Mantell (and possibly the rest of his audience at the British Museum) by describing cursorily a paper that he said he had had no time to finish, but then sent to the Athenaeum, which published it immediately. 'How sad and contemptible!' Mantell noted.[16] Even the usually calm Lyell took against Owen, who had attacked him in a review of Lyell's 'Address' in the *Quarterly*.

Mantell is recognised as a great palaeontologist, yet he has been unappreciated as a brilliant reporter of his times. His journal, for example, describes the death of Sir Robert Peel, the former prime minister, on 3 July 1850, after he was thrown from his horse while riding up Constitution Hill. According to Mantell, Peel was 'picked up senseless; the collar bone was fractured and severe concussion of the brain had taken place'. The horse seems to have fallen on him after he was thrown. Mantell expected him to recover; crowds packed Whitehall Gardens waiting for news. But the news was not good: 'Sir Robert Peel expired at eleven last night! This is a most deplorable event indeed . . . For several years Sir Robert has shown me every courtesy; invited me to dine with distinguished foreigners and savants; and sent me tickets for his conversaziones. He was the only eminent public man who payed any respect to art and science, apart from public policy.'[17]

The following year saw Mantell sharing in the exhilaration of the Great Exhibition in Hyde Park (even though Prince Albert had put Mantell's arch-antagonist, Richard Owen, in charge of the dinosaur exhibits). Mantell's own health continued to be poor; sciatica left him in pain for which he inhaled chloroform.[18]

By then he had moved, without his family, closer into London, to Chester Square in Belgravia, where he saw the bustling crowds heading for the Great Exhibition which covered nearly twenty acres of the park. For the opening, although very ill, Mantell walked with difficulty to Constitution Hill to watch Queen Victoria, Prince Albert and others process in the state carriage. When he finally got to the exhibition itself two days later, 'the interior surpassed all my expectations. It is quite overpowering. I cannot express the effect it has left upon my mind.' He also attended Prince Albert's opening of the Museum of Practical Geology on Jermyn Street (at which De la Beche gave the inaugural lecture).

Mantell pushed himself repeatedly to attend the exhibition, struggling against the pain in his spine and the press of the crowds – 30,000 on 4 October 1851, and then – with the closing deadline approaching – 97,000. He described seeing 'in the course of the day nearly 110,000 – one hundred and ten thousand! Vulgar, ignorant, country people: many dirty women with their infants were sitting on the seats giving suck with their breasts uncovered, beneath the lovely female figures of the sculptor. Oh! how I wished I had the power to petrify the living, and animate the marble: perhaps a time will come when this fantasy will be realised, and the human breed be succeeded by finer forms and lovelier features, than the world now dreams of.'[19] What also caught his eye, apart from the slovenly crowds, were exhibits of French machinery, Californian gold in quartz rock, and Austrian opal.

The Great Exhibition ended on Saturday 11 October 1851, with organs playing the national anthem, followed by the peals of bells, gongs 'and all manner of hideous and crushing sounds . . . no order was observed for the closing scene of the most marvellous display the world ever beheld!'[20] The sense of a new world of technical marvels was felt on 13 November when the telegraphic connection between Calais and Dover was completed. Mantell wrote in his journal of 'messages transmitted across the bottom of the Channel!!!! – by and bye this mode of communication will reach from one pole to the other I have no doubt'.[21]

Mantell died the following year, on 10 November 1852, nearly seven years before Darwin spelled out the theory and facts of evolution. Some attribute his death to suicide; others to an overdose of opium taken after falling on the way upstairs and crawling to bed. The theories are not contradictory. He had relied heavily on opiates since his accident. He did not live to see the Crystal Palace re-erected in south London in Sydenham. On New Year's Eve in 1853 a dinner was held inside what was to be the reconstructed *Iguanodon*, with Mantell's nemesis, Owen, at the head of the table.[22] Nor was he able to witness Owen's greatest contribution to British public and scientific life – the establishment in 1881 of the Natural History Museum.

Nonetheless, Mantell ended his life fully aware that what the famous Robert Bakewell, author of *An Introduction to Geology*, had told him in September 1829, was right: 'He says I shall ride on the back of my *Iguanodon* into the temple of Immortality!'[23]

So he did. Mantell's discovery confirmed what Cuvier had suggested earlier. There had been an age of reptiles preceding the age of mammals, and the dinosaurs had walked on land – some (as every child now knows) – like *Tyrannosaurus rex*, on two feet.

9

CELIBACY GALORE

With its High Church tradition weaker than Oxford's, the University of Cambridge advanced rapidly in science in the eighteenth and nineteenth centuries. When Cambridge's three-part examination for the bachelor of arts degree, known as 'the tripos', was reformed in the 1750s, mathematics became the basis of the new system. It was honoured as a support for the 'argument from design'. At Oxford, in contrast, the Greek and Latin classics were seen as central; mathematics and physics secondary.

Early in the Lent term of 1818, the news swept Cambridge that the Reverend John Hailstone, Woodwardian Professor of Geology for thirty years, was proposing to marry. A job vacancy thus opened. Under the stern provisions of the will of John Woodward, who had endowed the chair in 1728, the holder must be unmarried. Were he subsequently to wed, commanded Woodward from beyond the grave, 'his election shall be thereby immediately made void, lest the care of a wife and children should take the Lecturer too much from study, and the care of the Lecture'.

Woodward was a self-made man and amateur geologist whose life was changed in Gloucestershire in the 1720s when he discovered shellfish lodged in solid rock as well as beds of shellfish in ploughed agricultural land. He took these marine fossils to be evidence of Noah's Flood, in which he was a passionate believer, just as he was convinced that God had sent the Flood to punish the sinful human race and to recreate the earth to be 'more nearly accommodated to the present frailties of its Inhabitants'. Woodward attributed the source of the Flood to waters rising from beneath the earth's crust and submerging the entire globe, leaving shells and rock strata as they subsided.

At his death Woodward left his personal estate and effects to Cambridge University, which had granted him a doctorate in medicine. His astonishing collection of fossils was bequeathed to the university museum. Woodward was one of the early enthusiasts for whom collecting was part of geology's appeal. Paying others to assist him, he had gathered and catalogued 9,000 specimens which he housed in specially made walnut cases. So proud was he of his fossil collection that in his will he dictated that, for three days a week, the Woodwardian lecturer (who was to be paid £100 a year) should be present at the museum from nine until eleven o'clock in the morning and from two until four in the afternoon, 'to show the said Fossils gratis, to all curious and intelligent persons as shall desire a view of them for their information and instruction'.[1] His instructions did not stop there. The lecturer 'must be always present when they are shown, and take care that none be mutilated or lost'.

When Adam Sedgwick – a fellow of Trinity College and lecturer in theology and mathematics – applied for the Woodwardian post in 1818, he had no intention of mastering the science of geology. All he promised was to deliver public lectures 'on some subjects connected with the Theory of the Earth'. Like many others, Sedgwick was drawn to geology by poor general health. In 1812 he had broken a blood vessel while on a river excursion. Then, when one of his usual colds tended to a violent cough followed by inflammation of the lungs, his family and friends feared he would become consumptive. Indeed, he felt the effects of this chest illness throughout his life and pronounced himself 'unfit for sedentary labour after 1812'.[2]

The conviction that fresh air and regular exercise were indispensable determined him to become a candidate for the Woodwardian professorship. Looking back, he recalled to a friend how much he had enjoyed hunting as a young man: 'I was a keen sportsman till I became a professed Geologist. So soon as I was seated in the Woodwardian Chair I gave away my dogs and my gun, and my hammer broke my trigger.'[3]

In the contest for the chair, Sedgwick had one rival – G. C. Gorman of Queens' College – but, as Sedgwick later recalled, 'he had not the slightest chance against me, for I knew absolutely nothing of geology, whereas he knew a good deal – but it was all wrong!' (The editors of Sedgwick's collected letters assert that 'precedents were not wanting at Cambridge for the election of a man of ability to a Professorship in a subject of which he knew nothing'.[4]) While the contest for the post was under way, Sedgwick declared: 'Hitherto I have never turned a stone; henceforth I will leave no stone unturned.'[5] If William Buckland was a humorist, Adam Sedgwick was a wit. Among a geological generation of prose stylists, as a writer he was supreme. His letters are brilliant, narrative and descriptive. Sedgwick would become one of the great geologists of his era by inadvertence.

In 1818, at the age of thirty-three, Sedgwick became Woodwardian Professor of Geology. He would hold the post for the next fifty-five years.

As a poor Yorkshire vicar's son, Sedgwick had entered Cambridge in 1801 as a sizar – a term used for students at Cambridge and Trinity College Dublin who received some form of assistance, such as lower fees or lodging, and were thereby marked as persons of limited means. The study of divinity formed part of Sedgwick's programme, along with mathematics. He received his Bachelor of Arts in 1808, having come fifth in the top rank of mathematicians, known in Cambridge as 'Wranglers'. He then prepared for his fellowship examination. He accepted that he was entering a life of celibacy. Trinity College's stern ecclesiastical statutes proclaimed that all the fellows, save two, should be in priests' orders within seven years from the full completion of the degree of Master of Arts under threat of forfeiting their fellowship.

After his studies Sedgwick had wanted to see the Continent, which had been so long cut off by the Napoleonic Wars. But in the summer of 1816 he was able to spend four months in France, Switzerland, Germany and Holland. Sedgwick found the travel fiercely uncomfortable, with six people often crammed into a coach. His enjoyment, moreover, was hampered by his violent anti-Catholicism and dislike of the French. While he thought Paris 'a noble capital', he found the people 'so abominable and detestable that there can be no peace for Europe if they are not chained down as slaves, or exterminated as wild beasts'.[6]

Sedgwick met his clerical obligations in 1817 when he was ordained deacon by Bishop Bathurst of Norwich. Formally installed, he was pleased that 'on Monday the first, [I] commenced Resident in my own house . . . Our Residence is severe while it lasts. We are not permitted to be away from our houses for a single night.'[7] Attending service regularly, and preaching generally once each Sunday, he was sure, would wear him out. But by 1834 he had risen to higher clerical status, as prebendary of Norwich Cathedral.[8]

His clerical duties, sixty miles northeast of Cambridge, did not prevent him from continuing to lecture at the university. (He also became university proctor and headed the vice squad on Cambridge's streets, arresting fifteen women in one month alone for prostitution.) In 1819 Sedgwick delivered the first of his course of talks that made him the most popular lecturer at Cambridge. He continued these until 1872 when he was compelled by age and poor health to appoint a deputy. A witty and riveting speaker, he could hold an audience rapt for over an hour. His lectures were open to women, and when speaking he was often guarded. He told a patron: 'Geology introduces some tender topics which require delicate handling. I must speak truth, but by all means avoid offence if I can.'[9] (He did not say what these topics were.)

The same year he formed, with the distinguished geologist, John Henslow, the Cambridge Philosophical Society, 'for the purpose of promoting Scientific Enquiries, and of facilitating the communication of facts connected with the advancement of Philosophy'. The society's annual dinner was one of Sedgwick's red-letter days.

In November 1818, shortly after becoming Woodwardian professor, Sedgwick was elected fellow of the Geological Society. He welcomed the 'robust, joyous and independent spirits' he met at the GeolSoc and admired the way its members 'toiled well in the field, and who did battle and cuffed opinions with much spirit and great good will'.[10] He soon formed a close alliance with one of the society's younger members, Roderick Murchison. Murchison later observed that he was instantly drawn to the professor by his 'buoyant and cheerful nature, as well as from his flow of soul and eloquence'.[11] As a geological novice Murchison valued Sedgwick as much for his knowledge as for his friendliness – to the extent that Murchison was teased for hero-worship. He signed his letters '*Rodericus*'; Sedgwick in turn signed his 'Yours to the earth's centre'.[12] Their first expedition together was to Scotland, with the purpose of working out the clear relation of the red sandstones. Sedgwick found the variety of rocks astonishing. 'Arran,' he wrote, 'is a geological epitome of the whole world, and is, moreover, eminently picturesque.'[13]

While excavating at Robin Hood's Bay, near Scarborough, in 1821, Sedgwick suffered an unfortunate accident. As he described it to a friend: 'I have nearly lost the use of one eye in my combats with the rocks. A splinter struck it with such violence that it has for the last three or four days been of very little use to me.'[14] His eyesight never recovered. Thereafter he would struggle to work by candlelight.

In 1822 Sedgwick travelled to the northwest to explore the intricate geology of the Lake District. There he formed a close friendship with William Wordsworth – at whose house he was made welcome, yet who, as Sedgwick was well aware, had uttered 'a poetic ban against my brethren of the hammer'.[15] Sedgwick especially enjoyed their walks. 'Some of the happiest summers of my life were passed among the Cumbrian mountains, and some of the brightest days of those summers were spent in your society and guidance,' he told the poet many years later.[16] He wrote several letters headed 'On the Geology of the Lake District', which were included in Wordsworth's *A Complete Guide to the Lakes, comprising Minute Directions for the Tourist, with Mr. Wordsworth's Description of the Scenery of the Country, etc. And Three Letters on the Geology of the Lake District, by the Rev. Professor Sedgwick*.

The two men certainly did not agree on geological vocabulary. Admiring the cliffs of the coast of Somerset, Sedgwick had once written: 'They afford fine specimens of the contortions exhibited by that rock to which geologists have given the name of greywacke. What a delightfully sounding word! It must needs make you in love with my subject.'[17]

At a meeting of the British Association for the Advancement of Science, however, Sedgwick had fought hard against the introduction of the general term 'scientist', which had been proposed by William Whewell, theologian and historian of science. When Whewell argued at the meeting that 'we already have such words as "economist", "artist" and "atheist"', Sedgwick exploded: 'better die of this want [of a term] than bestialise our tongue by such a barbarism'.[18] Sedgwick lost the battle; the *Oxford English Dictionary* recognised the term 'scientist' by 1840.

The 'British Ass', otherwise more politely known as the 'BA', was continuing to grow. In his report for the association's first meeting, William Conybeare had emphasised that the study of fossils rather than of minerals was the cause of the success of English geology and that England was fortunate in having a great variety of rocks of 'the secondary series of formations: in these the zoological features of the organic remains associated in the several strata, afford characters far more interesting in themselves and important in the conclusions to which they lead than the mineral contents of the primitive series'.[19] The geological historian Nicolaas Rupke has called this summary of the contributions by the English school 'more than an exercise in the history and philosophy of science; it was an expression of chauvinism and of pride in the participation of geology in national progress, reform, and expansion of the Empire'.[20] He attributed English success to the fact that the extensive middle part of the geological record, the Secondary, 'belongs in a very considerable measure to England . . . an island blessed with a uniquely condensed and yet distinct series of most fossiliferous rocks'.[21]

By the time of the third meeting, in Cambridge in June 1833, the British Association for the Advancement of Science was clearly having national impact. The rising stars in early Victorian science attended, including Michael Faraday, Sir John Herschel, John Dalton, Charles Babbage, Sir David Brewster, Sedgwick, Whewell, Thomas Chalmers, Thomas Malthus and William Somerville. The association's pattern was to rotate its meetings around the major cities, always avoiding London. Successive venues were Edinburgh, 1834; Dublin, 1835; Bristol, 1836; Liverpool, 1837; Newcastle, 1838; Birmingham, 1839; and Glasgow, 1840. By this time more than two thousand people attended the annual meeting, the press coverage was huge, and the official membership was over a thousand.

In 1821 Sedgwick published a *Syllabus of a Course of Lectures on Geology* for his Cambridge students. The order of the older, Palaeozoic, rocks had not been yet worked out – ten years later he would be the first to define it – but he gave a good classification of sedimentary rocks. He emphasised the connection of geology with mineralogy. In his second edition, in 1832, he emphasised their separateness. On 3 March 1827, his letter to a fellow cleric announced: 'I was made vice-president of the London Geological Society at the last annual meeting. But this honour brings no grist. There is no manger in my stall, so that notwithstanding my V.P.G.S. at the tail of my signature, I may die of hunger.'[22]

He did not die, but rose to the presidency of the society within two years. Whewell wrote him a letter endorsed 'To be opened immediately' – an indication of Sedgwick's habitual carelessness with his correspondence. Sedgwick's eloquence served the society exceedingly well. In his opening presidential address in 1830, he declared: 'Each succeeding year places in a stronger point of view the importance of organic remains, when we attempt to trace the various periods and revolutions in the history of the globe.'[23]

A year later, his presidential term finished, he launched the subsequently influential Wollaston Medal, made possible by a Dr Thomas Hyde Wollaston, who gave £1,000 to the Geological Society, 'the income from which was to be used to promote researches concerning the mineral structures of the earth, or in rewarding those by whom such researches may hereafter be made'.[24] Wollaston died two weeks later.

This was the occasion when Sedgwick, awarding the medal to William Smith, the mapmaker, called him 'the Father of English Geology',[25] after which he got down to his main address, in which he made some highly important declarations. Lengthily and ringingly, he praised Charles Lyell for distancing himself in his *Principles of Geology* from Jean-Baptiste Lamarck, as well as from the doctrines of spontaneous generation and the transmutation of species 'with all their monstrous consequences'.[26]

In the next breath, Sedgwick attacked Lyell for arguing his own doctrine of uniformity as if he were a barrister with a brief. In Sedgwick's opinion, Lyell's 'uniformitarianism' implied that life had always existed on earth and that the cataclysms of the past were no different from the earthquakes, eruptions and tidal waves of their own day. That was not at all how Sedgwick saw the planet's development. He believed in repeated catastrophes. Twisted mountain layers and giant out-of-place boulders showed him that the earth had seen episodes of 'feverish spasmodic energy'.

More importantly, in the same speech, Sedgwick recanted what he dubbed a 'philosophic heresy' – that the biblical deluge could be equated with the evidence of watery cataclysm in the rock record. He declared that the vast masses of diluvial gravel scattered over the surface of the earth could not be attributed to 'one violent and transitory period'. He deplored the erroneous induction which had led many excellent observers of a former century to refer all the secondary formations of geology to the Noachian deluge. 'Having been myself a believer, and, to the best of my power, a propagator of what I now regard as philosophic heresy, I think it right,' he ended, 'as one of my last acts before I quit this Chair, thus publicly to read my recantation.'[27] James Secord describes Sedgwick's recantation of the deluge as Lyell's 'greatest theological triumph'.[28]

One of Sedgwick's geology students at Cambridge was the young Charles Darwin, who entered the university in January 1828.

In 1831 the pair went on a field trip together to North Wales to trace the juncture of limestone cliffs and Old Red Sandstone. (The Old Red, long a favourite subject of investigation for British geologists, is a thick sequence of sedimentary rocks named after the colour given by iron oxide, formed in the Devonian period, roughly 359 to 419 million years ago.) They set off from 'The Mount', Darwin's family home in Shrewsbury, where Charles's sister Susan was captivated by her brother's witty, urbane bachelor professor friend. The first day of their expedition in Snowdonia was nearly ruined, however, by Sedgwick's gloomy mood. 'I know that the d----d fellow never gave her the sixpence. I'll go back at once . . . ,' he suddenly cursed, referring to a tip he had left with the waiter at their hotel in Conwy to be given to their chambermaid. Darwin had to restrain the professor from walking all the way back across the mountains to check for himself.[29] Sedgwick, then forty-six, tramped heavy-laden, with a big iron hammer and a heavy leather collecting bag full of rocks.

When Darwin accepted the invitation to go on a round-the-world voyage on the HMS *Beagle*, Sedgwick recommended books for him to take. These included Charles Daubeny's *A Description of Active and Extinct Volcanos*, J. F. D'Aubuisson's *An Account of the Basalts of Saxony*, and Robert Bakewell's *An Introduction to Geology*. (He may have known that Darwin had already acquired Lyell's *Principles*.)

Having been on the same side in 'the Great Devonian controversy' (against De la Beche's claim to have found fossils of large plants in ancient strata that Murchison argued were too old to hold flora) did not prevent a serious break between Sedgwick and Roderick Murchison occurring in 1852; they reached an impasse in defining and naming the oldest rocks.

Twenty years earlier, in 1832, Sedgwick had rambled about various parts of North Wales, describing himself to a friend as 'burnt as brown as a pack-saddle, and a little thin from excessive fatigue'.[30] For scenery he decided he preferred the Lake District: 'The Welsh are a kind-hearted, but rather dull set of people; just made to be beaten by the Saxons. It is, however, wrong to judge of a people whose language one does not speak.'[31] But he was to spend more time in Wales on an important task. During the summer of 1834, he and Murchison – at that time his friend, and protégé – had spent four weeks on the Welsh Borders trying to set a boundary between Sedgwick's ancient rocks, which he named 'Cambrian' (after Cambria, the Latin name for Wales derived from 'Cymru', the Welsh word for Wales) and the somewhat younger rocks that Murchison had identified by distinctive fossils and called 'Silurian' (after the Silures, an ancient British tribe who lived in southeastern Wales at the time of the Romans).

In 1835 the men presented a paper to the Geological Society called 'On the Silurian and Cambrian Systems, Exhibiting the

Order in which the Older Sedimentary Strata Succeeded Each Other in England and Wales'. But their collaboration was doomed. In 1852 Sedgwick and Murchison fell into bitter conflict over the division between the Cambrian and Silurian. The heart of their dispute was the question of where the base of the Silurian stood in relation to the Cambrian. Murchison had claimed (and De la Beche had accordingly coloured on a geological map) part of Sedgwick's Cambrian as Silurian. On 25 February 1852, Sedgwick read a paper to the Geological Society in which he protested strongly against this insult from 'my friend and fellow-labourer, in this instance my antagonist'.[32] The ever-eloquent Sedgwick compared himself to a man who comes home to find 'that a neighbour has turned out his furniture, taken possession, and locked the door upon him'. In short, Murchison had 'Silurianized the map of Wales'. Their estrangement lasted for almost twenty years – only to be healed by Sedgwick's letter of condolence to Murchison on the death of his wife.

Sedgwick's own unmarried state irritated him his whole life. He was painfully conscious of it. As a young man in Cambridge he wrote to the Reverend W. Ainger, curate of Hackney: 'I wish some blooming damsel could contrive to kindle a flame in my breast, for then I might stand some chance of keeping up a proper degree of animal heat.'[33] He wrote to another clerical friend, whose wife was nursing him through an illness: 'If I had a wife I would sham ill now and then in order that she might make a pet of me.'[34] While he thought himself ugly, women were often drawn to him. A portrait of Sedgwick shows a tall handsome man with bright eyes, a strong chin and sensual lips. In 1837, Lyell wrote to his sister Eleanor that in Norwich it had been said of Sedgwick that he was 'so popular, with the ladies in particular', yet his observer commented, 'I hardly think he will ever marry now.'[35]

Inviting married fellow cleric Dr William Somerville and his wife – the celebrated science writer, Mary Somerville – to visit him at Trinity in April 1832, Sedgwick wrote 'we shall have a small party to welcome you and Mrs Somerville . . . A four-posted bed (a thing utterly out of our regular monastic system) will rear its head for you and Madame in the chamber immediately under my own; and your handmaid may safely rest her bones in a small inner chamber.'[36] The visit lasted a week.

Sedgwick's envy of happy marriages extended to the throne. When Queen Victoria and Prince Albert visited Cambridge in October 1843, Sedgwick wrote to his sister: 'it is plain that the royal pair love one another'.[37] Sedgwick and the prince consort got along very well and shared many views about science and university reform. The professor of geology's great contribution came in helping Albert reform the curriculum of Cambridge University. Unsurprisingly, Sedgwick was rewarded with several royal invitations.

On a visit in December 1847 to Osborne, Queen Victoria's country house on the Isle of Wight, bought and renovated by Prince Albert, Sedgwick told his sister (in an extremely long Christmas letter), how amused he was to see 'carts, toys and other signs of young children, scattered about the hall in considerable confusion'. Everything had 'a happy, domestic look'. Dressing for dinner was another matter. The gentlemen dressed like gentlemen in a private party – with one exception: 'They wear not loose trousers at Court, but tight pantaloons or shorts; of course I had, as a clergyman, a pair of shorts. I wish this custom had continued in society. For more than twenty years after I came to College men commonly appeared at dinner in shorts and silks.' He also observed that a dish of 'well-toasted oat-cake is handed round with the cheese. I believe the Queen likes it.' Sedgwick admired Prince Albert's knowledge and also his children: 'Five finer and more healthy children I never saw.'[38]

In his fifth decade, Sedgwick declined to accept the living of East Farleigh in Kent, worth an annual £1,000, because it would mean abandoning his Cambridge geology professorship. Yet it would have allowed him to marry. His decision was deplored by his friends, Lyell especially. Happily married himself, Lyell was convinced that were Sedgwick to leave Cambridge and marry 'he would be much happier, and would eventually do much more for geology'.[39] But it was never to be.

10

FROM SILURIA TO THE MOON

Roderick Impey Murchison would prove to be the most politically powerful of the 'brethren of the hammer'. Regarded by some as the greatest geologist of them all, Murchison was certainly the most combative. He turned geology into an instrument of British imperialism, directing skills into mapping and the exploration of countries within Britain's expanding empire. In 1844 his predictions of gold in Australia led to a gold rush. His encouragement of ventures such as the search for a Northwest Passage through Canada from the Atlantic to the Pacific is recognised today in at least fifteen features around the globe bearing his name. These include Murchison Sound in British Columbia, Murchison Falls in Uganda, Murchison River in Western Australia along with two of its tributaries (the Roderick and the Impey) and the Murchison River in New Zealand. There is also a Murchison crater on the moon.

Another mark of his successful cultural imperialism can be found in the British names given to the periods of the geological timetable. It was Murchison who decided to keep the basic categories of eras – Primary, Secondary, Tertiary – the huge blocks of time designated earlier by Abraham Werner. Into these he introduced important subdivisions that he called 'periods'. These were the Silurian, the Devonian and the Permian.

The Silurian was Murchison's most famous coinage. Its adoption was encouraged by the addition of a new name for the period between the Silurian and the Carboniferous – the Old Red Sandstone became the Devonian. As a military man, Murchison admired the ancient British tribe of Silures for their resistance to their Roman occupiers. As he himself explained: 'British geologists, therefore, will not doubt that "Siluria" is a name entitled to be revived, when they are reminded that these struggles of their ancestors took place upon the very hills which it is proposed to illustrate under the term "Silurian system".'

Though a relative latecomer to geology, Murchison ascended swiftly to the top. His first biographer, Archibald Geikie, writing in 1875, said: 'Such was the state of geological science at the time that a great work could be done by a man with a quick eye, a good judgment, a clear notion of what had already been accomplished, and a stout pair of legs.'[1] Murchison had all of these. He took up geology in 1823 – only eight years after William Smith's formulation of the law of strata identified by fossils. Despite his late start, he rose to rank above Buckland, Sedgwick, De la Beche, Conybeare and perhaps even Lyell.

Murchison was born in Scotland in 1792 into a wealthy Ross and Cromarty family where he acquired his love of fox-hunting. He attended Durham School and after passing out from the Royal Military College at Great Marlow spent eight years in the army. As captain with the 6th Dragoons he served in Spain and Portugal as the British tried to help the Spanish expel the invading French. In January 1809 he survived (as 900 others did not) the retreat from Corunna in northwest Spain. His military ambitions were scotched by the advent of peace in 1815.

Unlike many of his fellow ex-officers, Murchison avoided becoming a clergyman. On leaving the army he married Charlotte Hugonin of Nursted House, Hampshire, a general's daughter and woman of considerable intelligence and means. They toured Europe for the first two years of their marriage and settled initially in Barnard Castle, County Durham. It was there that Murchison met Sir Humphry Davy, who told him he was wasting his time fox-hunting. Davy encouraged him to come to London and throw himself into science. As a lure, Davy promised to get him into the Royal Society.

Murchison was persuaded. He saw geology as a science that would allow him free air and could be combined with his hunting and shooting. When in London, he began attending lectures on chemistry at the Royal Institution and in 1825 he joined the Geological Society. He loved its boisterous meetings and welcomed the acquaintance of Sedgwick, Conybeare, Buckland and Lyell. Within two years he was made a fellow. That summer, exploring with his wife, Murchison studied the geology of the south of England and concentrated on stratigraphical research. It was on the Jurassic Coast that he first met fossil-hunter Mary Anning (who, as we have seen, was as impressed by his good looks and soldierly bearing as by his knowledge). Murchison clearly enjoyed the experience, describing it as 'perhaps about the happiest period of my life'. That research provided the basis for his first scientific paper, read to the GeolSoc later in the year.

His explorations continued, accompanying Lyell to southern France, northern Italy, Switzerland and the volcanic region of Auvergne; then Sedgwick, to the Alps, resulting in their first joint paper. These field trips were no idle strolls through the countryside. They involved strenuous bending, hammering rocks, picking up and examining, then carrying them home. For twenty years Murchison undertook exhausting journeys in search of new strata. He was not a theoretician, rather an observer and reporter.

When geology was new, naming was as important as finding. To place the strata in accurate order of their deposition and to give a new division a name was an absorbing and fraught exercise. What were the identifying marks of their rocks – colour, texture, fossils or absence thereof?

In the 1830s, with Adam Sedgwick, Murchison began to study the Old Red Sandstone formations in the Welsh Borders – the long north–south strip between Shropshire and what is now the county of Powys. Murchison wondered whether the fossiliferous rocks underlying the Old Red Sandstone could be grouped into an order of succession.

It was a good question. The rocks and quarries near Builth Wells showed an astonishing frequency of trilobites. Murchison observed and described five species of these distinctive tiny creatures, with their domed eyes, fringed haloes, three lobes (whence the 'tri'), furrows and ribs.

His 1831 field trip – while Sedgwick was conducting his own investigations in North Wales – marked the occasion when, claims Martin Rudwick, Murchison came out from under Charles Lyell's shadow.[2] What Murchison looked back on as his 'Eureka moment!' came at a spot about eight miles east of Builth Wells, where a stream flows into the River Wye at Trericket Mill near Llanstephan Bridge. At this spot he saw the junction between the Old Red Sandstone and the Transition strata below. ('Transition' is a term formerly applied to the lowest uncrystalline stratified rocks – greywacke – supposed to contain no fossils. Such rocks were called 'transition' because they were thought to have been formed when the earth was passing from an uninhabitable to a habitable state.)

Murchison, repeating what he had already published, placed the Silurian above the Cambrian. Elaborating his classifications, he divided the Silurian into Upper (subdivided into Ludlow and Wenlock) and Lower (with subdivisions Caradoc and Llandeilo); and below the Lower Silurian, he (or with Sedgwick in a joint paper) listed the Cambrian – Upper Cambrian, Middle Cambrian and Lower Cambrian. Sedgwick then outlined his ideas on the lower rocks. Noting the absence of fossils the further down the Cambrian one looked, Sedgwick suggested that their first appearance marked the beginning of life itself. He was wrong because life had indeed emerged much earlier, but the fossils were hard to find. The sudden emergence of many varied fossils at the base of the Cambrian rocks is now referred to as the 'Cambrian explosion'.

In the Wye and also in the hills on the north bank of the river, Murchison could see slabs of greywacke, while in and around a small waterfall on the river's south bank lay the Old Red Sandstone, with its distinct fossils. According to Rudwick, Murchison at that moment 'found a project that might make him independently a leading man of science'. This gave him 'a distinctive personal stake in fossil-based methods in geology, and ranged him even more clearly behind his adopted father-figure, William Smith'.[3]

The outcome was Murchison's declaration of the Silurian system: a series of formations, each with distinctive organic remains, distinct from the older Cambrian rocks below. In his research notes from the same trip, Murchison drew sketches of other Welsh rocks, from the Swansea coalfields to the Transition rocks at Llandeilo – no small matter in the 1830s. He described the coalfields and overlying formations in South Wales and the English borders. His work in turn provided important new information for the burgeoning coal industry.

With his growing reputation, Murchison became president of the Geological Society in February 1832. Re-elected a year later, he was returned for a further term from 1841 to 1843. Lyell dedicated the third edition of his *Principles* to him.

In August 1835, Murchison and Sedgwick presented a joint paper to the British Association for the Advancement of Science meeting in Dublin on 'the Transition formations' in Wales. Murchison then tucked all these findings into 768 pages, published in 1839 as *The Silurian System* and dedicated 'to my dear Sedgwick'. Its publication marked the point when his reputation, if not his book sales, exceeded Lyell's.

It was only to be a few years, however, before Murchison and Sedgwick fell into bitter conflict over the division between Cambrian and Silurian. Neither man denied that Sedgwick had identified the Cambrian; but neither recognised the large overlap between the two. Murchison proved to be a formidable adversary, publishing in 1843 a geological map of Britain in which the whole of Wales was shown as Silurian. His designation of the Silurian system brought him the Copley Medal from the Royal Society in 1849, and drew a wordy accolade: 'for eminent services rendered to geological science during many years of active observation in several parts of Europe and especially for the establishment of that classification of the older Palaeozoic deposits designated the Silurian System, as set forth in the two works entitled *The Silurian System as founded on Geological Researches in England*, and *The Geology of Russia in Europe and the Ural Mountains*.'

In 1854 the publication of his book *Siluria*, which summarised Palaeozoic research across the globe with tantalising hints of where gold and coal might be found, secured for him, according to his biographer Robert Stafford, the 'reputation as the premier practical geologist of his day'.[4]

Murchison might be remembered even more for another designation: 'Permian'. This naming came about in 1840 when he went with two French scientists to carry out the first geological survey of European Russia. His trip, subsidised by the Emperor Nicholas I, took in the city of Perm in the Urals and prompted him to give the name 'Permian' to the last period of the Palaeozoic era. For his efforts, the emperor awarded him the Grand Cross of the Order of St Stanislaus and a ceremonial dagger (it sold at Bonham's in London for £48,000 in 2010).

In 1846, two years before Lyell received the honour, Murchison was knighted. A more unique title was awarded him during the 1849 meeting of the BA in Birmingham. Sir Roderick led an excursion to the nearby Dudley Caverns and up to the top of the Wren's Nest, a formation of exceptional Silurian features, and there invited the public to hear his lecture on the submarine formation of the rock on which they were standing. For this exercise he wore a green Tyrolean hat and a shepherd's plaid scarf, and bowed his head as the Bishop of Oxford dubbed him the 'King of Siluria'. (He used this pseudo-enthronement to seek a peerage through the Earl of Clarendon, an important member of the Whig government, but failed.)

He had more immediate matters on his mind, though. In 1855 he succeeded Henry (now Sir Henry) De la Beche, his old antagonist, as director general of the British Geological Survey and director of the Royal School of Mines, as well as of the School of Practical Geology. Moreover, he became one of the founders and subsequently four times president of the exploratory, Empire-minded, Royal Geographical Society. Murchison supported the African explorations of the Scottish medical missionary Dr David Livingstone, who discovered, among other places, the Victoria Falls. When Livingstone lost contact with the outside world for six years, Murchison led the Royal Geographical Society's campaign to try to find him.

Sadly, Livingstone's rescue by H. M. Stanley (who uttered the celebrated understatement, 'Dr Livingstone, I presume') came too late to be witnessed by Sir Roderick, who died five days before the discovery, on 22 October 1871, aged seventy-nine.

The loss of Charlotte two years earlier had hit Murchison badly, and he suffered a stroke in 1870 from which he only partially recovered. Her death had brought him, after twenty years' silence, a letter from Sedgwick, who apologised for having waited several months after his bereavement before writing. 'I did not wish to intrude on your sorrows too soon,'[5] he wrote. He hoped that Murchison could bear the loss of his wife 'like a Christian'.

It was not until both Murchison and Sedgwick were dead that the Cambrian–Silurian boundary dispute was resolved. In 1879 the distinguished geologist Charles Lapworth provided the solution. He introduced a new system, the Ordovician, to come between the Cambrian and Silurian and to cover the period he calculated as extending from 444 to 485 million years ago. Once again the Welsh past provided a name: the Ordovices were a tribe who had inhabited the area between that held by the Silures to the south and by the Cambrians to north, the region where the distinctive rock strata occur. The tripartite nomenclature – Cambrian, Ordovician and Silurian – became the standard. The Ordovician was subdivided into six units, starting from the lowest: Tremadoc, Arenig, Llanvirn, Llandeilo, Caradoc and Ashgill – the first four after place names in Wales.

Yet there has been increasing discontent with this provincial terminology. Some of the criteria used for identification of the base of each subdivision do not seem relevant in regions far from Britain and sometimes not even in Britain itself. The International Commission on Stratigraphy's subcommission on the Ordovician, after visits to many different rock sections throughout the world, proposed a new set of stages by which the bases could be widely recognised and correlated. After much discussion, most of the British names for these bases were dropped and new, more varied, names were introduced – including Floian, Dapingian and Katian.

Not everyone was pleased. 'What have they done to the Ordovician?' wailed John C. W. Cope, professor of palaeontology and stratigraphy at Cardiff University.[6] The answer seems to be that 'they' have dragged the terminology of the subdivisions into the twenty-first century and into the world of international geology. But the classic names for the systems – Cambrian, Ordovician and Silurian – remain.

11

ALPS ON ALPS ARISE

Rocks presented themselves not only as objects of study but also as peaks for conquest. Early geology demanded as much in cartography, draughtsmanship, physical courage and good luck as it did scientific method and deduction. Rock samples had to be gathered, sites visited, sketches drawn in situ and in all these pursuits dangers were encountered and overcome. None was more dramatic than the attempt by Edward Whymper, a young English illustrator and explorer, to conquer the Matterhorn in 1865. The Matterhorn was, and remains, the quintessential pinnacle of rock. An ideal peak drawn from collective imagination would look just like this Swiss pointed triangle; a Jungian archetype towering above Zermatt on the Swiss side and Cervinia on the Italian.

At 5.30 in the morning of 13 July 1865, Whymper, accompanied by two guides, Michel Croz and Peter Taugwalder, Taugwalder's two sons, and the alpinists, Lord Francis Douglas, the Reverend Charles Hudson and his nineteen-year-old friend Douglas Hadow (a novice climber) set out from Zermatt to climb the Matterhorn. It was Whymper's seventh attempt at the most difficult Alp, which had yet to be conquered. What happened over the next twenty-four hours is the stuff of climbing lore and legend.

The party arrived at the chapel at Schwarzsee, a village in south-western Switzerland, at 7.30 a.m. They left at 8.20 a.m. and an hour later stopped to calculate their route. After an hour's delay and fifty minutes' climbing, they arrived on the Hörnli Ridge. By half past eleven they had reached 9,900 feet. They made camp higher up on a solid platform just level with the Furggengrat, which curves northwest above the glacier that bears its name. The guides in advance had done some reconnaissance on the upper slopes and returned at three that afternoon with good reports of the snow slopes above them. The weather was calm as they camped there that night. As Whymper recalled:

> We passed the remaining hours of daylight – some basking in the sunshine, some sketching or collecting; and when the sun went down, giving, as it departed, a glorious promise for the morrow, we returned to the tent to arrange for the night. Hudson made tea, I coffee, and we then retired each one to his blanket bag; the Taugwalders, Lord Francis Douglas, and myself occupying the tent, the others remaining, by preference, outside. Longer after dusk the cliffs above echoed with our laughter and with the songs of the guides, before we were happy that night in camp, and feared no evil.[1]

The party moved on at first light.

In fact, first light is rather late in the day to start an Alpine ascent because snow and ice tend to melt and become unstable during the relatively high temperatures in full sun during the day. The likelihood of rock and icefall is greatly increased as the temperature rises above freezing. Alpine climbers must also consider seriously the time of their descent from the summit, a venture best done in daylight. Whymper's party included some of the finest mountaineers of a generation. Francis Douglas (brother of the Marquess of Queensbury) had only the week before achieved the difficult ascent of the Gabelhorn (13,363 feet). Charles Hudson, the vicar of Skillington in Lincolnshire, was a recognised Alpine climber who had climbed Mont Blanc ten years before – without guides – on a demanding route. His young friend Hadow was a formidable neophyte, having climbed Mont Blanc that year – his first Alpine climb – in under four and a half hours, returning to Chamonix after a five-hour descent. The feat would be impressive now, even with modern equipment.

The party was not roped together and there was little to hold on to. By 6.20 a.m., led by Hudson the vicar, moving over the snow slopes, they had reached 12,800 feet; by 9.55 a.m., when they stopped for fifty minutes, they had reached 14,000 feet, close to the northeast ridge.

At 1.40 p.m. the party reached the Matterhorn's summit: 14,782 feet. They paused for an hour or so to congratulate each other. They made a flag out of a guide's shirt and shouted down to a competing team of Italian climbers, 1,250 feet below the summit on the South West Ridge, hurling rocks at the group to get their attention. Whymper paid homage to the view from the top:

> Mountains 50 – nay a hundred – miles off, looked sharp and clear . . . details – ridge and crag, snow and glacier, stood out with faultless definition. Pleasant thoughts of happy days in bygone years came up unbidden as we recognised, the old family forms . . . Ten thousand feet beneath us were the green fields of Zermatt, dotted with chalets, from which blue smoke rose lazily . . . There were the most rugged forms, and the most graceful outlines – bold, perpendicular cliffs, and gentle undulating slopes; rocky mountains and snow mountains . . . There was every combination that the world can give, and in every contrast that the heart could desire.[2]

Twenty-five-year-old Whymper wrote with the Romantic sensibility of the age. In *The Prelude* in 1850, William Wordsworth wrote of looking up at Mont Blanc:

> That very day,
> From a bare ridge we also first beheld
> Unveiled the summit of Mont Blanc, and grieved
> To have a soulless image on the eye
> That had usurped upon a living thought
> That never more could be.[3]

Even earlier, in 1802, Coleridge wrote a 'Hymn Before Sun-Rise in the Vale of Chamouni' (his spelling of Chamonix) and appended a long note beginning:

> Chamouni is one of the highest mountain valleys of the Barouny of Facingny in the Savoy Alps; and exhibits a kind of fairy world, in which the wildest of appearances (I had almost said horrors) of Nature alternate with the softest and most beautiful. The chain of Mont Blanc is its boundary; and besides the Arve [river] it is filled with sounds from the Arveiron [another river with its source at Mont Blanc] which rushes from the melted glaciers, like a giant, mad with joy, from a dungeon, and forms other torrents of snow-water, having their rise in the glaciers which slope down into the valley.[4]

Preparing for the dangerous descent, Whymper and Hudson had agreed that their guide Croz would lead the party down, followed by Hadow, Hudson, Douglas, and Peter Taugwalder senior. Then as a separate pair would follow Whymper and Peter Taugwalder junior (his brother having returned to Zermatt before the summit attempt). Whymper wrote down the names of the party and placed them in a bottle at the summit. The group roped themselves together and set off down the mountain.

Down in Zermatt a boy ran into the Monte Rosa Hotel and reported seeing an avalanche falling from the top of the Matterhorn on to the Matterhorn Glacier 4,000 feet below the summit. What he had seen in fact was a disaster that occurred as Whymper's team, reaching the most difficult part of the descent, had paused because of the steepness of the slope. In Whymper's much quoted words:

> [Croz] was in the act of turning round, to go down a step or two himself; at this moment Mr. Hadow slipped, fell against him, and knocked him over. I heard one startled exclamation

> from Croz, then saw him and Mr. Hadow flying downwards; in another moment Hudson was dragged from his steps, and Lord Francis Douglas immediately after him. All this was the work of a moment. Immediately we heard Croz's exclamation, old Peter [Taugwalder] and I planted ourselves as firmly as the rocks would permit: the rope was taut between us, and the jerk came on us both as one man. We held; but the rope broke midway between Taugwalder and Lord Francis Douglas. For a few seconds we saw our unfortunate companions sliding downwards on their backs, and spreading out their hands, endeavouring to save themselves. They passed from our sight uninjured, disappeared one by one, and fell from precipice to precipice on the Matterhorn Glacier below, a distance of nearly 4,000 feet in height.[5]

For the next two hours Whymper and Taugwalder plus son made their way cautiously down, unnerved and shaken. The survivors descended following their original line of ascent.

The vivid description of how four men – Croz, Hadow, Douglas and Hudson – were lost appeared in Whymper's letter to *The Times*, 8 August 1865, giving the world an account of the most famous mountaineering accident of the nineteenth century. The letter concluded with a description of an unearthly vision that challenged the scientific world of the 1860s:

> When, lo! a mighty arch appeared, rising about the Lyskamm, high into the sky. Pale, colourless, and noiseless, but perfectly sharp and defined, except where it was lost in the clouds, this and an earthly apparition seemed like a vision from another world; and, almost appalled, we watched with amazement the gradual development of two vast crosses, one on either side . . . It was a fearful and wonderful sight; unique in my experience, and impressive beyond description, coming at such a moment.[6]

'Here then,' according to the science historian Andrew St George, 'was everything one could ask for from the individualistic strain of nineteenth-century thought and action. A group of talented amateur individuals pit themselves against sublime and raw nature; they gain the summit spectacularly, and some of their party fall to earth thousands of feet below; and to round off the most famous mountaineering accident of the century there was a celestial vision to tease the new scientific worlds of the 1860s. Rocks were always throwing up new challenges.'[7]

In August of the following year, 1866, Whymper returned to the Alps imbued with the new geological sensibility of the times. He determined to measure, record, analyse and explain the snow layers and snow temperatures at the summit of the Col de Valpelline, just north of Aosta. In one year he had ceased to think of the mountain landscape as something to be feared and it became something to be analysed and quantified.

As the Romantic view lost out to the scientific view, the arts lost out to the sciences. No one cared about an individual perception but rather sought knowledge that could be exchanged as data and, as in scientific practice, duplicated and replicated. However, in a review for the Royal Institution in 1892, the lecturer John Tyndall recalled the sense of sadness that overwhelmed him when he first saw the Matterhorn, in the summer of 1868. He turned his thoughts to the time when the mountain was 'more mountainous, more savage' and 'stronger' than it now seemed. Holding on to this feeling, he continued: 'Nor did thought halt there, but wandered on through molten worlds to that nebulous haze which philosophers have regarded, and with good reason, as the proximate source of all things. I tried to look at this universal cloud, containing within itself the prediction of all that has since occurred. Did that formless fog contain potentially the sadness with which I regarded the Matterhorn? Did the thought which now ran back to it simply return to its native home?'[8]

Whymper's experience on the Matterhorn expressed an imaginative engagement with the world which was nonetheless impersonal.

The novelty was neither an invention nor a discovery but rather a growth of a new way of thinking, a mode of thought that swiftly spread through the academic world. The work of palaeontologists, glaciologists and archaeologists in the nineteenth century was central to the arrival of scientific change in the form of four giant intellectual leaps forward.

The first was the idea of a field of physical activity occurring everywhere, even in vacuums. The second was the realisation that all matter is atomic. Indeed, by 1840, cell theory in biology and atomic theory in chemistry were established. Between the two, the nineteenth century established that electromagnetic effects – such as light waves – occurred within a continuous field and recognised ordinary matter as atomic.

The third big scientific idea of the nineteenth century was the conservation of energy. The fourth was the theory of evolution. These last two have had much to do with how we account for change or transition on a micro and macro scale, either during the short space of a chemical reaction or during the long time of evolutionary development. It was almost by chance that the theory of evolution developed in biology, for, at the start of the nineteenth century, the idea had been touched on by Immanuel Kant and Pierre-Simon Laplace. Laplace applied Newton's law of gravitation to the solar system (in his five-volume *Celestial Mechanics*, in which he applied Newton's law of gravitation to the solar system; he was also responsible for early probability theory).

John Tyndall, with his retrospective wisdom, commented in 1892 that Kant and Laplace had concluded the various bodies of the solar system had once formed part of a single mass and that, as the ages rolled away, the planets were detached and the chief portion of the hot cloud compressed and formed our sun. He found in the earth evidence of fiery origin as it had the same substances as the sun.[9]

The search for physical origins gave rise to the search for biological origins. The first people to question the perceived timescale of man's origins were Lyell, with his three books of *Principles* on matter, the

Scottish publisher and natural philosopher, Robert Chambers, and the archaeologist Boucher de Perthes, who first discovered stone tools and flint weapons in the gravels of the Somme River in 1832.

To intellectuals of the early nineteenth century, the superiority of the sciences was clear: here was the way forward. In the analysis of John Stuart Mill: 'no one dares to stand up against the scientific world, until he too has qualified himself to be named as a man of science; and no one does this without being forced, by irresistible evidence, to adopt the received opinion. The physical sciences, therefore (speaking of them generally), are continually growing, but never changing: in every age they receive indeed mighty improvements, but for them the age of transition is past.'[10]

Perhaps the most influential convert to the scientific approach through geology was Herbert Spencer. His search for a general theory was said by T. H. Huxley to be 'the first attempt to deal, on scientific principles, with modern scientific facts and speculations'.[11] What Spencer did was far more important than escorting George Eliot to the theatre.

Born in 1820, not dying until 1903, this gifted man led his century in persuading his contemporaries that the scientific approach was the rational way to look at physical phenomena. As agnostic, philosopher, geologist, economist, psychologist, phrenologist and iconoclast (and, it is claimed, inventor of the forerunner of the paperclip, as well as of the phrase the 'survival of the fittest'), Spencer unified thinking on the physical world and, under the influence of Eliot and her partner George Henry Lewes, wrote the nine-volume *System of Synthetic Philosophy*. He led others in coming to his strong belief in the unity of scientific knowledge – his century's great contribution to human philosophy.

12

DARWIN THE GEOLOGIST

While Charles Darwin is inescapably identified with biology, his enduring scientific love was always geology. Arriving at a new place and trying to puzzle out its past from its rocks and fossils was, he wrote to his sister in May 1832, more fun than 'the first day's partridge shooting'. Nothing could compare 'to finding a fine group of fossil bones, which tell their story of former times with almost a living tongue'.[1] Having begun his journey round the world on the *Beagle* in December 1831, he found that he could hardly sleep at night for thinking about the geological phenomena he was seeing.

As a sixteen-year-old enrolled to study medicine at the University of Edinburgh, Darwin thought he had dismissed geology for ever. He abandoned his medical studies because he could not stand the blood and suffering he witnessed. On visiting an operating theatre he saw what he described as 'two very bad operations, one on a child . . . I rushed away before they were completed. Nor did I ever attend again, for hardly any inducement would have been strong enough to make me do so; this being long before the blessed days of chloroform.' He also gave up geology. Its lectures were so dull, he recalled in his autobiography, that he vowed 'never so long as I lived to read a book on Geology or in any way to study the science'.[2]

He changed his mind, however, in January 1828 when he moved to Cambridge and fell under the spell of the brilliant Adam Sedgwick. Three years of study culminated in 1831 in their joint field trip to Wales to trace the junction of limestone cliffs and the Old Red Sandstone. Like Sedgwick, Darwin did not travel light.

A young acquaintance, Robert Lowe, who accompanied him part of the way, remembered that Darwin 'carried with him, in addition to his other burdens, a hammer that weighed 14 pounds'. So much did Lowe admire Darwin that 'I walked twenty-two miles with him . . . a thing which I never did for anyone else before or since.'[3]

On returning from Wales, Darwin learned that his Cambridge mentor and friend, the botanist John Henslow, had recommended him to go as intellectual companion to Captain Robert FitzRoy on a journey to circumnavigate the globe. Darwin had the additional advantage of being a naturalist, although in fact there was an official naturalist on board the *Beagle*.

On the voyage Darwin brought an alert curiosity to his reading of Lyell's *Principles*, the book Captain FitzRoy had shrewdly given to him before they left England. FitzRoy was acquainted with Lyell, who had asked him to look out for several features when he got to South America.

After an infuriating succession of delays, the *Beagle* eventually sailed from Plymouth on 27 December 1831. Its first stop came in January at the island of St Jago (São Tiago) in the Cape Verde islands, west of the bulge of North Africa. The tropical lushness of the island dazzled Darwin. He had never seen anything like its jungle, birds, insects, flowers and volcanic rocks. His powerful impressions of the island were used by his great-great-granddaughter, the poet Ruth Padel, in her 2009 poem 'Like Giving to a Blind Man Eyes' – written for the 200th anniversary of Darwin's birth:

> Lava must once have streamed on the sea-floor here,
> baking shells to white hard rock. Then a subterranean force
> pushed everything up to make a new island.
> Vegetation he's never seen, and every step a new surprise.
> 'New insects, fluttering about still newer flowers. It has been
> for me a glorious day, like giving to a blind man eyes.'[4]

For Darwin, St Jago provided, apart from his brief time in Snowdonia, his first field research. Guided by Lyell's *Principles*, he looked at nature as an example of processes still continuing. On St Jago he became 'convinced of the infinite superiority of Lyell's views over those advocated in any other work known to me'. He noticed a horizontal band of white rocks filled with seashells and coral standing thirty feet above the ground. There was only one possible explanation: a gradual uplifting of the land, a process that Lyell – and before him his eighteenth-century predecessor, James Hutton – had described. To Darwin the larger meaning was unmistakeable: the earth was ancient.

All his life Darwin would credit Lyell for his awakening on that island. As he wrote in his *Autobiography*:

> The geology of St Jago is very striking yet simple: a stream of lava formerly flowed over the bed of the sea, formed of triturated recent shells and corals, which it baked into a hard white rock. Since then the whole island has been upheaved. But the line of white rock revealed to me a new and important fact, namely that there had been afterwards a subsidence round the craters, which had since been in action, and had poured forth lava. It then first dawned on me that I might write a book on the geology of the countries visited, and this made me thrill with delight. That was a memorable hour to me . . .[5]

He was so much under Lyell's influence that he likened the elevated white rock to Lyell's Temple of Serapis at Pozzuoli. He also wrote home and asked to be proposed for membership of the Geological Society.

Darwin's *Beagle* notebooks are heavily weighted with his geological observations. As he travelled he collected rocks and fossils; he packed an immense box of bones, shells and fossils and sent even larger boxes back to John Henslow in Cambridge. Much has been made in recent years of Darwin's observations on the varied species of finches and mocking birds on the Galápagos Islands. Less noticed is the fact that his notebooks contain more about rocks than about living creatures. The half dozen large islands, which lie on the Equator, are volcanic in origin and Darwin wanted 'a good look at an active Volcano', as he told Henslow.[6] In fact, he saw a great many and walked to the tops whenever possible.

The *Beagle* notebooks also contain what is only now being brought to public attention: Darwin's hatred of slavery. Coming from an abolitionist background (both his grandfathers, Erasmus Darwin and Josiah Wedgwood, were outspoken campaigners against slavery), Darwin was horrified in Brazil to see that the practice was even crueller than he had imagined: 'Every individual who has the glory of having exerted himself on the subject of slavery may rely on it [that] his labours are exerted against miseries perhaps even greater than he imagines.'[7]

Towards the end of its voyage, in April 1835, the *Beagle* sailed northwards along the coast of Chile. There Darwin saw the famed terraces of Coquimbo – flat parallel trails in the mountains high up over the port city. He took the sight as evidence of the recent emergence of the whole of the Chilean coastline and concluded that the elevation of the Andes was still continuing. It was while in Chile that he first experienced an earthquake. It shattered his sense of reality. 'The world, the very emblem of all that is solid,' as he later wrote, shuddered beneath his feet 'like a crust over a fluid'.[8]

Back from the *Beagle* – his home for nearly five years, from December 1831 to October 1836 – Darwin realised his true vocation was to be a geologist. He was immediately admitted to the Geological Society and was henceforth entitled to style himself 'FGS', although he thought it 'a great pity that these and the other letters, especially F.R.S. [Fellow of the Royal Society] are so very expensive'. Admission cost him six guineas and an annual three guineas. Yet his worth was appreciated; he would be on the society's council for the next thirteen years.

He met Charles Lyell, who suggested that Darwin might also like to apply to join the new Athenaeum Club: 'if you like to dine at the club do so. There is no vacancy, but you stand the first of those who are knocking at the door for admission.'[9] The club had been founded in 1824 by the then president of the Royal Society, Sir Humphry Davy, and the Tory politician John Wilson Croker, and was intended for literary, scientific and artistic gentlemen. It quickly acquired 900 members and Michael Faraday was its first secretary. Unusually for the time, the club did admit ladies, if only on Wednesday evenings. Some members grumbled at the invasion, Lyell observed in a letter to his sister: 'retreated into the library, which was respected at first, but now the women fill it every Wednesday evening, as well as the newspaper room, and seem to me to examine every corner with something of the curiosity with which we should like to pry into a harem. They all say it is too good for bachelors, and makes married men keep away from home.'[10]

After meeting Lyell in person, Darwin felt enormous admiration for him and they became fast friends. He wrote to his brother on 6 November 1836: 'Amongst the great scientific men, no one has been nearly so friendly and kind, as Lyell.'[11]

Lyell shared the appreciation for friendship; he told Darwin: 'I have spent the last week entirely in comparing recent shells with fossil Eocene species, identified by Deshayes. When some great principle is at stake, all the dryness of minute specific comparisons vanishes, but I heartily long for some one here with a collection of shells, and leisure to talk on these matters with.'[12] He advised Darwin: 'Don't accept any official scientific place, if you can avoid it, and tell no one that I gave you this advice, as they would all cry out against me as the preacher of anti-patriotic principles. I fought against the calamity of being President as long as I could.'[13]

At this time the accumulating fossil record was making the changes of species during the geological past the central question in geology. By March 1837 Darwin had concluded that species were not immutable; he opened his first notebook on the subject in the following month.

Two years later, in June 1838 – at the age of twenty-nine, unbearded and not yet married – Darwin took himself up to Scotland to see for himself the great geological conundrum of the Highlands, the so-called 'Parallel Roads of Glen Roy'. He was well aware that he knew little about British rocks.

His motives were not entirely geological. Overworked by writing his *Beagle* notebooks, he needed to get some fresh air for his health. He had promised his intended bride – his cousin Emma Wedgwood – that he would finish the Glen Roy paper and then they would marry.

Darwin had addressed the issue of marriage before he left for Scotland. Ought he to remain a bachelor? He posed the question to himself on a scrap of blue paper, drawing up two columns, one headed 'Not to Marry', the other 'Marry'. On the 'Not Marry' side, he set thirteen advantages such as 'Freedom to go where one liked – choice of Society & little of it. Conversation of clever men at clubs – Not forced to visit relatives, & to bed in any trifle. – to have the expense & anxiety of children – perhaps quarrelling. – Loss of time. – cannot read in the evenings – fatness & idleness – . . . if many children forced to gain one's bread' and 'Travel. Europe, yes? America????' Against these, in the 'Marry' column, he set ten advantages such as 'Children (if it please God)' and 'a nice soft wife on a sofa with good fire, & books & music perhaps'. It was no choice. He concluded emphatically: 'Marry – Mary [sic] – Marry Q.E.D.'[14]

On 23 June 1838 Darwin took a steamboat to Edinburgh and arrived five days before Victoria – who had just turned eighteen – was crowned queen. He walked round the city where he had briefly attended medical school. Next he made his way to Fort William on the west coast, then headed north, passing the distant snow-capped mountain of Ben Nevis until he reached the Lochaber region and the town of Spean (pronounced '*Spee-an*') Bridge. After a night's sleep, he entered the nearby valley of Glen Roy. At once he saw the puzzling natural configuration that he had come to explain (and which remains to be seen, in the same form, today).

The first written account of the parallel roads appeared in 1776, in the appendix to the third volume of a guidebook by Thomas Pennant called *A Tour of Scotland*: 'In the face of these hills, both sides of the glen, there are three roads at small distances from each other and directly opposite on each side . . . They are carried along the sides of the glen with the utmost regularity, nearly as exact as drawn with a line of rule and compass.'[15] Indeed, along both sides of the wide and beautiful glen run three horizontal parallel terraces – the so-called 'roads' – as flat as if they had been levelled by a giant grass roller. Not only flat but wide – sixty feet across. The lower shelf lies 200 feet below the middle, the topmost 100 feet above that. The valley itself encompassing these roads is triangular. Narrow at the northeastern end, at the southwestern end it is open wide. One obvious explanation to Darwin was that the shelves were remnants of a former lake. But if so, what had held the waters in? The answer, Darwin decided, was that the valley had not held a lake; it had been an inlet of the sea and the waters had risen and subsided as the sea level changed.

Darwin's hunch followed Lyell's own. In the fifth edition of *Principles* (the fourth edition of 2,000 copies having been nearly all sold in one year), Lyell had called the Glen Roy shelves marine beaches.[16] Darwin himself could hardly forget that a year before he had seen similar parallel terraces in Chile. Where the terraces were littered with seashells he had swiftly concluded that the roads were beaches of the Pacific Ocean, their parallelism showing him that the earth had risen at three distinct times. He drew detailed diagrams of the sight in his notebooks.

This dramatic and distant phenomenon was just the subject Darwin wanted for his debut at the Geological Society. When the occasion arrived, he delivered a paper on his proofs of the 'recent elevation of the coast of Chile'. No intelligent person, he said, could doubt the rise of the land or the falling of the sea. Since he, probably alone of their number, had actually seen Chile with his own eyes, no one contradicted him. How could Glen Roy be anything but the same? After eight days in the Scottish valley, Darwin was convinced that he was seeing the same phenomenon as he had observed in Chile. The shelves were old sea beaches that had risen as the land was lifted in stages above sea level.

However, there was one important difference. On the shelves of Glen Roy there were neither seashells nor barnacles. Their profusion had struck him at Coquimbo, where he interpreted them to have been deposited as the mountains rose in a series of stages from the sea. Darwin was not going to let that small detail bother him. He wrote to Lyell on 9 August 1838 how he had enjoyed 'five days of the most beautiful weather, with gorgeous sunsets & all nature looking as happy as I felt'. He was fully convinced '(after some doubting at first) that the shelves are sea-beaches, although I could not find a trace of a shell'. He thought he could 'explain away most, if not all of the difficulties'.

Returning home to Shrewsbury that summer and then shifting back to London, Darwin finished his paper on 6 September. 'Eight good days in Glen Roy' he put in his notebook. 'My Scotch expedition answered brilliantly.' He then began reading Thomas Malthus's *Essay on the Principle of Population* (first published forty years earlier), noting that: 'In October, 1838, that is, fifteen months after I had begun my systematic enquiry, I happened to read for amusement *Malthus on Population*. Here, then, I had at last got a theory by which to work.'[17] Malthus held that human beings would continue to increase until they surpassed the amount of food available to feed them and that the poorer would always lose out to the richer in the struggle for food. In other words, the population was increasing geometrically while the food supply increased arithmetically.

It was not just from an intellectual interest in population. His father, Robert Darwin, a doctor, had warned Charles that he should marry soon if he wanted healthy children. Darwin himself was drawn to what Malthus called the 'fruitfulness of marriage' and would soon himself provide a notable example.

The day after he became engaged in November 1838, Darwin confided in Lyell that he felt 'the most sincere love and hearty gratitude to her [Emma] for accepting such a one as myself . . . I hardly expected such good fortune would turn up for me'. He also told Lyell: 'I deeply feel your kindness and friendship towards me, which in truth, I may say has been one chief source of happiness to me ever since my return to England.'[18]

He did not spell out for Lyell, a religious man, how strongly he and Emma differed on Christianity, however. Emma had written to him upon their engagement to thank him for his openness in telling her his anti-religious opinions, 'but my own dear Charley we now do belong to each other & I cannot help but be open with you'.[19] Nourished by the hope of an afterlife, she feared that her Charley endangered his chances by 'casting off' what Jesus had done for his benefit and for the whole world.

In December 1838, Darwin sent the completed Glen Roy manuscript – nearly ninety handwritten pages – to Lyell who, as a fellow of the Royal Society, forwarded it with a covering letter to one of the official secretaries. Darwin hoped they would not ask him to shorten it because, he said, there was not a sentence he could leave out. His Glen Roy paper, however, was not read to members until the next month, and then only by the society's secretary who delivered it in three mumbled instalments.

Before its publication, Sedgwick read the paper and asked for a drastic shortening. Darwin refused and the paper went forward to publication in full. It ended: 'The conclusion is inevitable, that no hypothesis founded on the supposed existence of a sheet of water confined by barriers, that is, a lake, can be admitted as solving the problematical origin of the "parallel roads of Lochaber".'

It was Darwin's first significant scientific paper and secured his election as a fellow of the Royal Society later that month (on 24 January 1839). Not for another eight years did the Royal Society change its rules so that new members would be elected on the basis of their contribution to science rather than for their social position or wealth. (This reform, long overdue, was introduced by Lyell's Scottish geologist father-in-law, Leonard Horner, in 1846.[20]) But Darwin qualified on his own merits. He also paid £70 for the privilege. In choosing the senior institution to hear his Glen Roy paper, he had cold-shouldered the younger Geological Society. Five days after his election he married.

Darwin's own work now moved to the study of species and their formation. He paid little attention the following year to the theory of the naturalist Louis Agassiz, who produced the later incontrovertible argument that the roads of Glen Roy (and much of the landscape of Europe and North America) had been formed by glaciers and not by the action of the sea and crustal elevation, as Darwin had maintained.

Long after, in 1859, Darwin received the Geological Society's highest honour, the Wollaston Medal, for his outstanding contributions to geology. These were identified as three books drawn from his *Beagle* voyage: *The Structure and Distribution of Coral Reefs* (1842), *Volcanic Islands* (1844) and *Geological Observations on South America* (1846). His Glen Roy paper was not mentioned. Not until 1861 – two years after the publication of his world-shaking *On the Origin of Species* – did the young Scottish geologist Thomas Jamieson publish a more explicit and detailed reconstruction of how the glacial flow of ice had dammed lakes, thus creating the Parallel Roads of Glen Roy.

Darwin had to concede that his Glen Roy paper was wrong, telling Lyell: 'I am smashed to atoms about Glen Roy.'[21] Courteously, he wrote to Jamieson: 'Your arguments seem to me conclusive. I give up the ghost.'[22] He asked permission to forward the letter to Sir C. Lyell. (Darwin himself was never knighted.) But his Glen Roy error scarcely mattered. By then his name would never again be associated with rocks.

13

THE ICEMAN COMETH

In its third decade, the Geological Society of London met for dinner at the Crown & Anchor Tavern in the Strand before adjourning to the society's rooms in Somerset House. While members were dining on the evening of 7 June 1840, a little over a mile away an attempt was made to assassinate the young Queen Victoria and her new husband Prince Albert as they were being driven up Constitution Hill. As the diary of a GeolSoc member records: 'the news spread rapidly and naturally created a deep sensation'. The culprit was later declared insane and transported to Australia. The royal couple had been married only a few months.

For geologists, a more memorable event occurred three days later. The Swiss geologist Louis Agassiz read a paper titled: 'On the polished and striated surfaces of the rocks which form the beds of glaciers in the Alps'. In it Agassiz, of the University of Neuchâtel, offered the then-unimagined idea that much of Europe's northern half had once been covered by a vast sheet of ice, as indeed had much of the northern hemisphere.

The glacial hypothesis offered an explanation for the long scratches visible on the surface of many rocks. It also gave the first convincing answer to the mystery presented by huge boulders lying on utterly different kinds of rock – granite on limestone, for example – which showed that they were far from their place of origin. These conspicuous misfits were known to geologists as 'erratics' (or, to borrow a phrase from James Joyce's as-yet-unwritten *Ulysses*, 'Wandering Rocks'). Lyell, in the first edition of *Principles*, had suggested that drifting icebergs had carried the erratics around the landscape.

Three years before his London appearance, in an address to the Swiss Society of Natural History at Neuchâtel, Agassiz had announced the *Eiszeit* (Ice Age). He spelled out his discovery that, in the relatively recent past (that is, the Pleistocene era – a term later coined by Lyell), a massive glacier had covered Switzerland and central Europe, reaching the borders of the Mediterranean and Caspian Seas. He attributed his awakening to the observations made in 1760 by the first scientific explorer of the Alps, Horace-Bénédict de Saussure. De Saussure had pointed out that miles below the snouts, or front edges, of the glaciers, the rock surfaces were scratched and smoothed, in striking contrast to the frost-splintered peaks above. He concluded that the glaciers had formerly extended many miles beyond their present limits and were retreating. Agassiz agreed with de Saussure; from 1837 to 1845 he had spent every summer but one in the Swiss Alps observing the movement of glaciers. He even had a hut built on one glacier so that he could keep a close watch on it. In 1840 he published his *Etudes sur les glaciers*.

From that day the *Eiszeit* theory was associated with Agassiz's name. The concept usefully removed a global flood from geological speculation. It also brought Agassiz, between 1836 and 1840, grants totalling £240 from the Geological Society of London and the British Association for the Advancement of Science.

In November 1840 Agassiz came in person to deliver his theory of the recent Ice Age to the Geological Society. A handsome young man with a confident smile, he described glaciers as having existed not only in Switzerland, where they were yearly seen by thousands of foreign visitors, but in Ireland, northern England and Scotland. In his paper titled 'On Glaciers, and the evidence of their having once existed in Scotland, Ireland, and England', he suggested that glaciers had once extended much further than they did at the present day and had been the primary agents in gouging out the huge U-shaped valleys in Switzerland. These moving rivers of ice scratched the rocks as they slid over, carrying and dumping erratics which slipped down the slopes as the mountains rose. They also threw off debris on either side, forming ridges he termed 'lateral moraines'.

Agassiz included in his talk his own explanation for the Parallel Roads of Glen Roy. The 'roads', he said, were the remnants of glacial lakes – former shorelines not of the sea, but of a freshwater lake dammed by a glacier, thus directly contradicting Darwin's paper given two years earlier. Darwin was at the meeting, as was Sedgwick, who joined in the fiery debate which 'kept up till near midnight'. He later read Agassiz's study of glaciers, and found it 'excellent', though he felt in 'the last chapter he loses his balance, and runs away with the bit in his mouth'.[1]

Overall, Agassiz's theory solved many puzzles, not least the existence of large boulders totally unlike the rocks around them. Lyell had suggested that these erratics had been carried by floating icebergs, but Agassiz's explanation was more convincing, and answered other questions as well. Glaciers carried debris as well as boulders and deposited them as they crept slowly along, scratching the rocks underneath. (By the ninth edition of *Principles*, Lyell would emphatically attribute the presence of erratic boulders on the Alpine peaks to glacial actions.[2])

Buckland, then president of the society, endorsed him wholeheartedly. Turning the chair of the meeting over to George Bellas Greenough, he told of his own conversion. He described himself as having been a 'sturdy opponent of Professor Agassiz when he first broached the glacial theory', but, after going to the Alps, he had been persuaded. Buckland then read the first part of his memoir on the evidence of glaciers in Scotland and the north of England and spelled out his own view: the Glen Roy roads had been made by successive inland lakes, dammed by a glacier. Ever the joker, Buckland condemned those who questioned the validity of scratches, grooves and polished surfaces of the glacial mountains 'to be damned to the pains of eternal itch without scratching'.[3]

Agassiz was challenged by Greenough, the founding president of the Geological Society: 'Does Prof. Agassiz suppose that the Lake of Geneva was occupied by a glacier 3,000 feet thick?'

Agassiz shouted back: 'At least!'[4]

Greenough was unimpressed. The glacial theory, he said, was the 'climax of absurdity in geological opinions'. Murchison, for his part, complained about the poetic terms Agassiz had used to describe glacial phenomena. Buckland, however, defended Agassiz's choice of words. He said that Agassiz had done well to revive de Saussure's expressive phrase *roches moutonées* (rounded, elongated bedrocks resembling the backs of sheep) whose characteristic shape was attributed to glaciers passing over them.

Murchison was sarcastic. If ice were the explanation for marks on the surface of many rocks, 'the day will come when . . . Highgate Hill will be regarded as the seat of a glacier, & Hyde Park & Belgrave Square will be the scene of its influence'. (The joke was self-flattering, for all present knew that Murchison had recently moved into fashionable Belgrave Square.)

Early the next year, in his presidential address, Buckland summarised his conclusions on glaciers and in particular on the roads of Glen Roy: these had been caused by two glaciers descending from the mountain of Ben Nevis across the valley of the Spean.

In fact, after a few months, Agassiz's ideas were accepted. From Edinburgh the naturalist Edward Forbes wrote to the professor, now back at the University of Neuchâtel, 'You have made all the geologists glacier-mad here . . . they are turning Great Britain into an ice-house.'[5]

Louis Agassiz, born in 1807, had received the degree of doctor of medicine in Munich in 1830, but his fascination with extinct forms of fish had moved him into natural history. He went to Paris and studied geology with Alexander von Humboldt and zoology with

Cuvier. At the age of twenty-two he produced his first book (based on collections brought back by explorers) on the history of Brazilian fossil fish. Extending his interest to European varieties, he wrote a history of freshwater fish in central Europe and then became professor of natural history at Neuchâtel. He received grants for research from Frederick William III, King of Prussia. His fame spread with the publication in 1833 of the first volume of his *Recherches sur les poissons fossiles*.

Agassiz had first come to Britain in 1833 when Buckland invited him to study the remarkable fish fossils in the Old Red Sandstone. He met the Scottish geologist Hugh Miller, who would write five volumes on the subject. The following year he went to Lyme Regis and examined the fossil fish collections of geologists Elizabeth Philpot and Mary Anning. He named a species of fish after Anning: *Belenostomus anningiae*. In the final 1843 volume of his *Recherches sur les poissons fossiles*, he thanked both Philpot and Anning.

In Switzerland in 1836 he inspected the glaciers of the valleys of Chamonix and Diablerets and also the moraines of the Rhône Valley. There he heard his friend Jean de Charpentier advocate the glacial concept as an explanation for the moraines and erratic blocks in the Alpine regions of Switzerland.

Perhaps de Charpentier should have had the credit for discovering the Ice Age. Nonetheless, it was Agassiz who named it and who delivered his theory in an address to the Swiss Society of Natural History in 1837. This dramatic announcement – that in relatively recent history a massive glacier had covered Switzerland and central Europe – drew Buckland to Switzerland the following year to tell Agassiz that the same evidence could also be found in Scotland. Buckland's speculations opened Agassiz's imagination to the concept of a giant ice sheet: not a separate one for Switzerland and another for Scotland, but one huge blanket of ice. The theory was welcome to Buckland, for it swept away any remaining delusions of deluges. Floods of ice, not floods of water, had shaped the landscape.

In the late summer of 1840, Agassiz came to Scotland for a third visit, this time for the Glasgow meeting of the British Association for the Advancement of Science. Arriving in August, Agassiz toured the Highlands with Buckland and had no difficulty in finding examples of striations, glacial polish, erratics, *roches moutonées* and moraines just as he had seen in Switzerland. Buckland then led him to the puzzling Parallel Roads of Glen Roy. Gripped by the spectacle, Agassiz wrote his own description – 'around the walls of the Glen Roy valley run three terraces, one above the other, at different heights, like so many roads artificially cut in the sides of the valley, and indeed they go by the name of the "parallel roads"'[6] – and concluded that they were the result of glacial action. He did not keep his theory to himself: on 3 October 1840, from Fort Augustus, he wrote to the influential daily newspaper the *Scotsman*, which four days later published his letter announcing his solution to the problem of the Parallel Roads of Glen Roy:

> at the foot of Ben Nevis, and in the principal valleys, I discovered the most distinct moraines and polished rocky surfaces, just as in the valleys of the Swiss Alps, in the region of existing glaciers; so that the existence of glaciers in Scotland at early periods can no longer be doubted. The parallel roads of Glen Roy are intimately connected with this former occurrence of glaciers, and have been caused by a glacier from Ben Nevis. The phenomenon must have been precisely analogous to the glacier-lakes of the Tyrol, and to the event that took place in the valley of Bagne.[7]

When Agassiz moved on to Ireland, Buckland went to visit Lyell on the family's Kinnordy estate and there converted him to Agassiz's glacial theory. Buckland reported back to Agassiz: 'Lyell has adopted your theory in toto!!!!' for 'solving a host of difficulties that have all his life embarrassed him'.[8]

Indeed Lyell had. He introduced new material into the 1840 edition of his *Principles*, writing in the preface: 'a chapter has been introduced for the first time on the power of river-ice, glaciers, and ice-bergs, to transport solid matter, and to polish and furrow the surface of rocks. The facts and illustrations contained in this chapter have been almost entirely derived from my private correspondence during the last four years, or from new publications.'[9] These must have included Agassiz's two-volume *Etudes sur les glaciers*, dedicated in part to Charpentier, published later that year.

In 1846 Agassiz left Britain – for good, as it turned out – when his name was put forward by Lyell to give the distinguished Lowell Institute lectures in Boston, Massachusetts. En route Agassiz's ship stopped at Halifax, Nova Scotia, and he was excited to observe 'the familiar signs, the polished surfaces, the furrows and scratches, the line engraving of the glacier . . . and I became convinced . . . that here also this great agent had been at work'.[10]

In America he discovered a country where the study of geology was well developed, and a Boston society that loved him and his lectures. Agassiz would stay on to become Harvard's first professor of natural history. There his teaching enthralled the young Henry Adams, who wrote in his autobiography that 'the only teaching that inspired his imagination was a course of lectures by Louis Agassiz on the Glacial Period and Palaeontology, which had more influence on his curiosity than the rest of the college instruction together'.[11]

Agassiz married a Bostonian (his first wife having died in Switzerland). Her name was Elizabeth Cabot Cary and in 1879 she became co-founder and first president of Radcliffe College, the women's liberal arts college (soon dubbed 'Harvard's Annex', now formally part of Harvard). In 1859, through his efforts, Harvard established the Museum of Comparative Zoology, which holds Agassiz Hall. He was the museum's first director.

Agassiz's good looks and urbanity, as well as his knowledge, were much admired in both Britain and the United States. In 1856 he received the Wollaston Medal from the Geological Society of London for his work on fossil ichthyology and in 1861 was awarded the Royal Society's illustrious Copley Medal. In New England the poet Henry Wadsworth Longfellow composed 'The fiftieth birthday of Agassiz' in his honour. Agassiz's own work now had American focus: a multi-volume *Natural History of the United States* was published between 1857 and 1862; he had completed only four of his planned ten volumes by his death in 1873. By then he was perhaps the most famous scientist in the world.

In recent years Agassiz has been much criticised for his theory that there were several human species, not just different races of a single species. But such ideas were not unusual at the time and should not be used retrospectively to cloud his great contributions to natural history. In 1884, another New England man of letters, Oliver Wendell Holmes, included Agassiz in a poem, 'At the Saturday Club'. The club had started in 1855 with monthly informal meetings at the Parker House in Boston. It drew together poets, scientists and philosophers to dine and enjoy good conversation. In Holmes's poetic reconstruction:

There, at the table's further end I see
In his old place our Poet's vis-à-vis,
The great PROFESSOR, strong, broad-shouldered, square,
In life's rich noontide, joyous, debonair.
His social hour no leaden care alloys,
His laugh rings loud and mirthful as a boy's –
That lusty laugh the Puritan forgot,–
What ear has heard it and remembers not?

. . .

How does vast Nature lead her living train
In ordered sequence through that spacious brain,
As in the primal hour when Adam named

The new-born tribes that young creation claimed!–
How will her realm be darkened, losing thee,
Her darling, whom we call our – AGASSIZ![12]

This poetic portrait of Agassiz appropriately included a reference to his greatest discovery – the slow creeping crawl of the glacier.

14

FOOTPRINTS IN PENNSYLVANIA

In 1841, five years before he put forward Agassiz's name, Charles Lyell had given his own series of lectures at the illustrious Lowell Institute in Boston. Lyell was delighted to have the opportunity of seeing America for the first time, and he would see a great deal: a thirteen-month trip, with his wife Mary, was projected. Even so, he was anguished at the thought of being parted from Darwin for a whole year.

The two men would not be London neighbours much longer. Darwin, now married, with a pregnant wife and increasingly unwell himself, was planning a move from their Bloomsbury home, Macaw Cottage, to a house in the country, in Downe, Kent. At first Charles and Emma had liked London. 'We are living a life of extreme quietness,' he told a friend, 'and if one is quiet in London, there is nothing like its quietness – there is a grandeur about its smoky fogs, and the dull distant sounds of cabs and coaches; in fact . . . I am becoming a thorough-paced Cockney.'[1] But with his weight dropping and energy failing, he now found London 'a vile and smoky place'.[2]

What would Lyell do without him? He moaned to Darwin in a letter: 'It will not happen easily that twice in one's life even in the large world of London a congenial soul so occupied with precisely the same pursuits, and with an independence enabling him to pursue them, will fall so nearly in my way.'[3] Whenever they saw each other, they could not stop talking geology. Darwin told Emma how the Lyells had called on him one morning after church: 'I was quite ashamed of myself today, for we talked for half-an-hour unsophisticated Geology, with poor Mrs. Lyell sitting by, a monument of patience. I want practice in ill-treating the female sex . . . few husbands seem to find it difficult to effect this.'[4]

With Mary, his wife of nine years, Lyell left from Liverpool in July 1841 on the steamship *Acadia*, which covered up to 250 miles per day. Casting his geologist's eye on the Atlantic, he observed that the ocean changed from deep blue to green as it became shallower over the Newfoundland banks. On 31 July, after eleven days, they reached Halifax, Nova Scotia, and posted the letters they had written on the voyage. Thirty hours more saw them arrive in Boston. This was six months before the much-anticipated visit of Charles Dickens. When Dickens landed on 3 January 1842, he was mobbed in a wave of adulation that would continue during the rest of his American visit.

While Lyell had no such celebrity, he was far from unknown. In 1838 he had published *Elements of Geology*, devoted to stratigraphy and palaeontology and their use in dating rock strata. The new book was very successful, appearing in subsequent editions, with much interchanging of material between it and the best-selling *Principles*.

Unlike his travels in France, where everything seemed to contrast with England, in the United States Lyell was struck by 'the resemblance of every thing I see and hear to things familiar at home'. The people especially, he thought, were 'so very English'.[5] (He wrote in 1841 before the great wave of Irish immigration began, after the start of the four-year potato famine.) It cannot have been long before Lyell saw the possibility of another bestseller. British interest in the country was intense and few visitors (emigrants apart) made the long ocean-crossing. Darwin, for one, had been no nearer America than the Galápagos.

Lyell was moved to recount his travels with the confidence of a reporter who knows that he is seeing what his readers will never see. 'One of the first peculiarities that must strike a foreigner in the United States is the deference paid universally to the [female] sex, without regard to station,' he noted. 'Women may travel alone here in stage coaches, steamboats, and railways, with less risk of encountering disagreeable behaviour, and of hearing coarse and unpleasant conversation, than in any country I have ever visited. The contrast in this respect between the Americans and the French is quite remarkable.'[6] He also observed that the social class divisions so apparent in England were of little interest here. When he asked the innkeeper to find his wagon driver, the landlord called out: 'Where is the gentleman that brought this man here?' Lyell was not accustomed to being called a 'man' while his employee was a 'gentleman'.[7]

Lyell noticed that the wildflowers and weeds were very different from those in England; yet two-thirds of the seashells were the same species as British shells: 'I shall have many opportunities of pointing out the geological bearing of this curious, and to me very unexpected fact.' He noted also the striations and polished surfaces – the signs of glaciers. But why, if Boston was on the latitude of Rome, were its winters so cold?

Before giving his series of lectures, he and Mary set out on a tour to see the geologist's dream: Niagara Falls. (Mary, having no children, was free to accompany him on his travels. Regrettably for today's reader, Lyell's copious travel notes omit any mention of what his wife wore on these arduous field trips.) In 1841 getting to western New York State was no simple matter. Travelling merely from Boston to New York City was more complicated than crossing the Atlantic. First, from Boston, the Lyells went west to Springfield, Massachusetts – 'by excellent railway', as he would write in his subsequent *Travels in North America; with Geological Observations on the United States, Canada and Nova Scotia*. They covered the first hundred miles 'in three hours and a half, for three dollars each'. They then 'descended the River Connecticut in a steamboat', stopping at

Hartford to see a large mass of vertical rock columns of red sandstone, then at New Haven.

After admiring the fine avenues, Yale University and the East and West Rocks that flank the city, the Lyells embarked on a large steamship which took them in less than six hours to New York City. Yet another steamship took them up the Hudson River, where on its west bank Lyell admired the columnar basalt called the Palisades. Reaching Albany, he was happy to meet state government geologists and sympathised with their dismay that, after $200,000 (or, as he put it, 40,000 guineas) had been expended on exploring New York State's mineral structure, they learned that no coal would ever be discovered nearby. 'This announcement caused no small disappointment, especially as the neighbouring state of Pennsylvania was very rich in coal.'[8]

Going by train from Albany to Niagara, Lyell was surprised to see 'one flourishing town after another, such as Utica, Syracuse, and Auburn' and a group of Oneida Indians offering for sale trinkets and moccasins of moose-deer skin and boxes of birch bark. Near Rochester, accompanied by the geological surveyor for northwestern New York, he saw the remains of a fossil mastodon and fragments of an ivory tusk. His main conclusion: the fossils of those North American rocks were not identical to those found in equivalent strata on the other side of the Atlantic.

In his lectures Lyell would make good use of his observations on the recession of Niagara Falls and the creation of the Niagara Gorge – a magnificent example, he declared, of geological forces at work. He had already described in *Principles* the history of the falls, going back nearly 10,000 years, and had calculated that, receding at about a foot a year, they were about 35,000 years old. Now with his own eyes he saw what he knew already: the mile-broad Niagara River flowed from Lake Erie into Lake Ontario. Before reaching the falls, the river was divided by an island which split it into two sheets of water that cascaded down with an enormous roar and pounded on the shale beds below so that their disintegration was constant.

Lyell endorsed the general opinion that the Falls would retreat to Lake Erie twenty-five miles away in 30,000 years.

Observing that the ridges above the table land between Lakes Ontario and Erie matched the ridges on the uplands bounding the valley of the Ottawa River, he wrote in his *Travels*: 'I shall content myself with stating that, with the exception of the parallel roads or shelves in Glen Roy, and some neighbouring glens of the Western Highlands in Scotland, I never saw so remarkable an example of banks, terraces, and accumulation of stratified gravel, sand, and clay, maintaining, over wide areas, so perfect a horizontality, as in this district north of Toronto.'[9]

When he and Mary went on to look at French Canada they encountered the French-English divide. English settlers with whom Lyell spoke described as 'enlightened' the measures introduced by the late Lord Sydenham (first Governor of Canada) to quash the French. One of Sydenham's supporters told Lyell: 'We shall never make any thing of Canada until we anglicise and protestantise it.' To this charge, Lyell heard 'a French seigneur' rejoin with bitterness: 'Had you not better finish Ireland first?'

Lyell concluded that Canada must continue to owe her protection from external aggression, not to local armaments and provincial demonstrations, but to the resources of the whole British Empire.

Heading south in the United States, he cast an envious eye on the great continuous coalfield of Pennsylvania, Virginia and Ohio. He saw the virtues of hard anthracite coal: nearly smokeless, it was much preferable to the soft bituminous coal used in London where it left people 'living constantly in a dark atmosphere of smoke, which destroys our furniture, dress, and gardens, blackens our public buildings, and renders cleanliness impossible'. Dickens, writing nine years later on the opening page of *Bleak House*, evoked the filth of the English capital: 'Smoke lowering down from chimney-pots, making a soft black drizzle, with flakes of soot in it as big as full-grown snow-flakes . . . Dogs indistinguishable in the mire . . . Fog everywhere.'[10] Why, Lyell wondered, were the British and American types of coal so different when both originated in the same decaying plant matter and came from the same species of plants formed at the same period? (In fact, the coals emerged from completely different types of plant and bituminous and anthracite represented these origins.)

While examining the coal seams in Pennsylvania, Lyell was fascinated by the use of gravity to get passengers down a hill from the Lehigh Summit Mine: 'we descended for nine miles on a railway impelled by our own weight, in a small car, at the rate of twenty miles an hour. A man sat in front checking our speed by a drag on the steeper declivities, and oiling the wheels without stopping.'[11] The coal came down the same way, with sixty mules ready to draw up the empty carts every day.

He found monotonous the long unbroken summits of the ridges of the Allegheny Mountains, but the scenery below beautiful. And he admired the religious tolerance and universal education – so unlike Great Britain 'where we allow one generation after another of the lower classes to grow up without being taught good morals, good behaviour, and the knowledge of things useful and ornamental because we cannot all agree as to the precise theological doctrines in which they are to be brought up'.[12] However, he had to get back to Boston. He described how he and Mary accomplished it:

> from Philadelphia by New York to Boston, 300 miles, without fatigue in twenty-four hours, by railway and steam-boat, having spent three hours in an hotel at New York, and sleeping soundly for six hours in the cabin of a commodious steam-ship as we passed through Long Island Sound. On getting out of the cars in the morning, we were ushered into a spacious saloon, where with 200 others we sat down to breakfast, and learnt with surprise, that, while thus agreeably employed, we had been carried rapidly in a large ferry-boat without perceiving any motion across a broad estuary to Providence in the State of Rhode Island.[13]

At some point during his New England trip Lyell discovered Martha's Vineyard. He was carried by 'an excellent railway' to New Bedford and found:

> a steam-boat in readiness, so that, having started long after sunrise, I was landed on 'the Vineyard', eighty miles distant from Boston, in time to traverse half the island, which is about twenty miles long from east to west, before sunset. Late in the evening I reached the lofty cliffs of Gayhead [Lyell's spelling] more than 200 feet high, at the western end of the island, where the highly-inclined tertiary strata are gaily-coloured, some consisting of bright red clays, others of white, yellow, and green sand, and some of black lignite . . . I collected many fossils here, assisted by some resident Indians, who are very intelligent.[14]

In October, he began his Lowell lectures. He admired Boston with the proximity of Harvard ('the best endowed university in America') and the stimulating Bostonian mix of professors, writers, lawyers and doctors. Their influence, Lyell felt, placed the state legislature 'under the immediate check of an enlightened public opinion'. He had been invited by Mr (John Amory) Lowell – a cousin of the founder – whom he described as 'trustee and director of a richly endowed literary and scientific institution in this city', with the instruction to deliver a 'course of twelve lectures on geology during the present autumn'.[15]

Geology was enjoying a great burst of popularity in the United States, with the Association of American Geologists being founded in 1840 in Philadelphia. Lyell's lectures were held in the Odeon Theatre, hired for the purpose. He noted wryly that while the public 'have gratuitous admission to these lectures by several judicious restrictions', such as requiring ticket applications several weeks in advance, the trustee 'has obviated much of the inconvenience arising from this privilege, for it is well known that a class which pays nothing is irregular and careless in its attendance'.[16]

Attendance was hardly irregular. Lyell was such an attraction that more than 3,000 people came to each lecture, with the consequence that he gave each twice, repeating the evening talk the next afternoon. 'Among my hearers were persons of both sexes,' he described, and 'of every station in society, from the most affluent and eminent in the various learned professions to the humblest mechanics, all well dressed and observing the utmost decorum.'

Travelling north of Boston to Lowell, in November 1845, Lyell inspected the mills which had made the Lowell family fortune. Once more what he saw contrasted with England. He admired the young women who worked the spinning-wheels; he found them: 'good-looking, and neatly dressed, chiefly the daughters of New England farmers, sometimes of the poorer clergy. They belong, therefore, to a very different class from our manufacturing population, and after remaining a few years in the factory, return to their homes, and usually marry.'[17] Factory work, he judged, was better for young women than domestic service 'as they can earn and save more; their moral character stands very high, and a girl is paid off, if the least doubt exists on that point. Boarding-houses, usually kept by widows, are attached to each mill, in which the operatives are required to board; the men and women being separate.'[18] He noticed also that few children were employed and that those under fifteen were compelled by law to go to school three months in the year under penalty of a heavy fine.

Lyell covered much ground. Moving south, he went to Washington, attended a debate in the Senate and met the new president, John Tyler. (Formerly vice-president, Tyler had become president in April 1841 after the new president William Henry Harrison died suddenly just one month after his inauguration.) He then crossed the Mason–Dixon line in order to examine the geology of the tertiary strata on the shores of the James River.

In Virginia he had his first sight of slaves, and was told by a white English immigrant from Hertfordshire that there was no room in the slave states for poor whites, as they were 'despised by the very negroes if they laboured with their own hands'.[19] He learned also of an arrival from New England, a man who, moving south to Virginia, had sold off the slaves and introduced Irish labourers, thinking they would be more economical. But after three years, so the New Englander told him, the Irish became very dissatisfied at being looked down upon by the whites as if they were black.

In January 1842, Lyell reached Georgia. Learning that the bones of a mastodon and other extinct mammals had been found, he decided that, at a comparatively recent period, the Atlantic had been inhabited by the existing species of marine testacea (shellfish) and that there had been an upheaval and laying dry of the ocean bed. This new land supported forests in which the megatherium, mylodon, mastodon, elephant, and a species of horse and other quadrupeds had lived and left remains that were found buried in the swamps. He later wrote: 'As no species of equus existed in the New World when it was discovered in the fifteenth century, naturalists were inclined, at first, to be incredulous in regard to the real antiquity of this fossil but as the tooth is more curved than in the recent horse, ass, or zebra, the fossil species may have differed as widely from any living representative of this genus, as the zebra or wild ass from the horse of Arabia.'[20]

At Savannah, Georgia, in order to see the bed of clay containing some mastodon bones, Lyell was instructed to be on the ground by daybreak and at low tide. Accordingly, he left in the middle of the night with his host's servant as a guide, 'and I found him provided with a passport, without which no slave can go out after dusk'.[21]

From his host, who spoke of 'abolitionists' in tones of loathing, Lyell heard that insurrection had been 'stirred up' by abolitionist missionaries. He himself took to heart his host's fear of the whites being outnumbered. As he wrote in his *Travels*: 'nearly half the population of Georgia are of the coloured race, who are said to be as excitable as they are ignorant. Many proprietors live with their wives and children quite isolated in the midst of the slaves, so that the danger of any popular movement is truly appalling.'[22] To Lyell, the black slaves appeared 'very cheerful and free from care, better fed than a large part of the labouring class' of Europe: 'We might even say that they labour with higher motives than the whites – a disinterested love of doing their duty.' Lyell was aware that there was another view but concluded that he 'found it impossible to feel a painful degree of commiseration for persons so exceedingly well satisfied with themselves'.[23]

Lyell's racism was not confined to people of colour. In New York, he remarked on 'the heterogeneous composition of its people . . . an endless procession of Irish parading the streets, with portraits of O'Connell emblazoned on their banners'. He observed that 'Pennsylvania also labours under the disadvantage of being jointly occupied by two races, those of British, and those of German extraction'. He asked himself whether it was 'dangerous to entrust every adult male with the right of voting. Yet in America they think the experiment a safe one, or even contend that it has succeeded. But some disagree.'[24]

Back in Washington he turned again to geological observation. He was surprised to see large masses of floating ice brought down from the Appalachian Hills. The huge boulders on the low grounds reminded him of how vast a territory in the South he had passed over without encountering a single erratic block. These far transported fragments of rock, he decided, were 'decidedly a northern phenomenon'.[25]

Moving up to Philadelphia he was shown the entire skeleton of a large fossil mastodon, 'or so-called Missourium', brought from Missouri and assembled, with some errors. As he noted in his *Travels*: 'This splendid fossil has since been purchased by the British Museum, taken to pieces in London, and correctly set up again under the direction of Mr. [Richard] Owen.'[26]

Returning to Boston in a hot July, the Lyells were delighted once more 'to see our friends, some of whom kindly came from their country residences to welcome us. Others we visited in Nahant, where they had retreated from the great heat, to enjoy the sea-breezes. Ice was as usual in abundance; the iceman calling as regularly at every house in the morning as the milkman. Pine-apples from the West Indies were selling in the streets in wheelbarrows. I bought one of good size, and ripe, for a shilling, which would have cost twelve shillings or more in London.'[27]

In October 1842 the Lyells returned to London, bringing with them thirty-six boxes of fossils. Soon after Charles would begin to write his *Travels in North America*, published in two volumes in 1845.

During this time back in Britain, Lyell was appointed by the Home Office as one of the experts charged with investigating the explosion at Haswell colliery, County Durham, which had killed ninety-five miners. The use of such high-profile scientists as Lyell and Michael Faraday (who was now Fullerian Professor of Chemistry at the Royal Institution) to investigate how such mining explosions occurred and to suggest steps towards prevention was unprecedented. Their subsequent recommendation to the Peel government was that firedamp should be drawn away from the mine by specially made conduits. Lyell's particular contribution, submitted on 21 October 1844, was that miners should be better educated, as the miners seemed unaware of the dangers of lighting their pipes from the flame of the Davy Lamp. These recommendations would lead to a major political row over the expense that implementing such proposals would entail.

Within two years Lyell was back again in Boston, delivering a series of Lowell lectures once more. In all, he would make four visits to America.

For Lyell and many since, the landscape of America was an open textbook on geological history. More's the pity that he never saw what were the greatest pages in the book – the Grand Canyon. As he had pronounced after his first visit: 'Certainly in no other country are these ancient strata developed on a grander scale, or more plentifully charged with fossils; and, as they are nearly horizontal, the order of their relative position is always clear and unequivocal.'[28]

In Kentucky he inspected a place of 'great geological celebrity' called Big Bone Lick, where the bones of mastodons and many other extinct quadrupeds had been dug up in extraordinary abundance. 'The term Lick is applied throughout North America to those marshy swamps,' he explained to his readers, 'where saline springs break out, and which are frequented by deer, buffalo, and other wild animals for the sake of the salt, whether dissolved in the water, or thrown down by evaporation in the summer season, so as to encrust the surface of the marsh. Cattle and wild beasts devour this incrustation greedily, and burrow into the clay impregnated with salt, in order to lick the mud.'[29]

Lyell's method of finding local specialists when abroad was to go to a pharmacy upon entering a new town and ask if anyone was interested in geology. He had discovered in Sicily in 1828 that the 'apothecary in the Pharmacia always set himself up as a savant'. In Jackson, Mississippi, this tactic brought down a physician who lived above the drugstore, who had read *Principles* and who had a collection of fossils.

Lyell saw the Mississippi River and its delta for the first time. He was unprepared for the French atmosphere of New Orleans, even less for the fossil teeth and jawbone of a mastodon shown him. Going up the Mississippi by steamboat, he was surprised when the booking clerk asked him if he wrote about geology and then questioned him about a recent article in the *Edinburgh Review*. Lyell was gratified with this evidence that the countries shared a common culture despite the wide ocean separating them. In Missouri in March 1846, Lyell saw the place where, thirty-five years before, an earthquake had altered the course of the Mississippi – an event he had already described in *Principles*.

Lyell gave more thought to the phenomenon of slavery during his travels. At first he had thought that the liberation of slaves was desirable, but changed his mind when he considered the obstacles. Who would feed the slaves if their masters stopped doing so? Where would they live? He decided, on balance, that he would not embrace the abolitionist cause. 'I have seen nothing to alter my views of the condition of the slaves,' he wrote to his father in February 1846 (having just sent off a long letter on the geology of Alabama to the Geological Society of London). 'If emancipated, they will suffer very much more than they will gain. They have separate houses, give parties, at which turkeys and all sorts of cakes are served up. They marry far more than our servants – eat pork – the women exempted from work a full month after childbirth, corporal punishment excessively rare; they do so much less bodily work than the whites in the North, that the Southern planters will not believe in the stories of the former.'[30]

He had already written to his father-in-law about the slaves in Georgia: 'My dear Horner . . . Besides being clothed and very well fed, they are by nature most peaceful, so that they hardly ever fight, and the contrast of some Irish labourers who came here to dig a canal (to which, by the way, we owe the discovery of the Megatherium), would really be laughable if it were not such a serious evil.'[31] In South Carolina he became conscious of the numerical preponderance of slaves. He sympathised with the reasoning behind the strict laws against importing books relating to emancipation and also with the prohibition on bringing back slaves who had been taken by their masters into free states.

Like Agassiz, Lyell wondered whether the black and white races had separate origins. Obviously influenced by his reading of Lamarck (although in *Principles* he had firmly denounced the Frenchman's ideas on the genetic transmission of acquired characteristics), he asked himself whether if 'the Negro' learned skills, he might be able to pass improved intelligence to his children. He wrote to his father-in-law about the differences in races between whites and 'people of colour' and raised the possibility that, with the passage of geological time, there might come about a being 'as much transcending the white man in intellect as the Caucasian race excels the chimpanzee'.[32]

When Darwin read Lyell's *Travels* in the summer of 1845 he was dismayed at his friend's treatment of slavery. 'Your slave discussion disturbed me much,' he wrote to Lyell, 'but as you would care no more for my opinion on this head than for the ashes of this letter, I will say nothing except that it gave me some sleepless, most uncomfortable hours.'[33] Darwin thought Lyell showed too little feeling and suspected that his view had been tempered by the hospitality he received from his anti-abolitionist hosts.

When his second trip to the United States began in 1845, Lyell had established that his English readers' appetite for American detail was virtually insatiable. In New York he found the most striking change to be the electric telegraph poles, standing thirty feet high, a hundred yards apart, 'and certainly not ornamental. Occasionally, where the trees interfere, the wires are made to cross the street diagonally.'[34] One result was the swift spread of international news: one item lay behind 'the sensation just created here . . . and is now going the round of the newspapers, namely, the conversion to the Romish Church of the Rev. Mr. Newman, of Oxford'.[35] He heard heated discussions of politics as people argued over the possible candidates for the next presidential election, to be held in 1848. He did not keep silent. 'I enlarged on the superior advantages of a hereditary monarchy, as preventing the recurrence of such dangerous agitation.'[36]

In New Jersey he made notes on the abundant dogwood in the forests 'with such a display of white flowers as to take the place of our hawthorn'. With greater experience, surveying the blue Appalachian mountain chain, with its many folds, arches and troughs, Lyell wrote a little homily for those still unaware of the age of the earth or, as he put it, 'not accustomed to reflect on the long succession of natural events . . . which have concurred to produce a single geological phenomenon, such as a mountain chain'. Although it appeared paradoxical, he attributed the structure of such mountains to 'the sinking, rather than to the forcing upwards, of a portion of the earth's crust'. The Appalachian chain, in short, was due to subsidence at a time when the mountains were still submerged beneath that ocean in which they were originally formed, as 'is testified by their imbedded corals and shells'.[37]

On seeing the confluence of three great navigable rivers, the Monongahela, the Allegheny and the Ohio, he marvelled at the sight of level seams of coal lying open on the river banks. He rhapsodised about the great future that lay waiting 'when the full value of this inexhaustible supply of cheap fuel can be appreciated; but the resources which it will one day afford to a region capable, by its agricultural produce alone, of supporting a large population, are truly magnificent'.[38] Moving on in his *Travels in North America*, Lyell called his readers' special attention to the vast Illinois coalfield, taking in parts of Illinois, Indiana, and Kentucky – an area almost as big as the whole of England.

Immigrants always caught Lyell's eye, although he was never tempted to become one himself. He observed 'many wagons of emigrants from Pennsylvania, of German origin' who were heading for the West. He spoke with an Irish grandmother who had lived in America for forty years but would 'die happy could she but once more see the Cove of Cork'.[39] He also asked his readers to abstain from drawing general conclusions from the conversation of persons whom chance had thrown in the traveller's way. Even so: 'As soon as we were recognised to be foreigners, we were usually asked whether we had made up our minds where we should settle. On our declaring that, much as we saw to like and admire in America, we had no intention of exchanging our own country for it, they expressed surprise that we had seen so many States, and had not yet decided where to settle. Nothing makes an English traveller feel so much at home as this common question.'[40]

Determined that on this trip he would devote an entire month to the geology of Nova Scotia, Lyell was delighted to find in the ancient rocks on the shores of the Bay of Fundy marks of ripples and raindrops. Lyell split the slabs, found more footprints on the underside and packed them to bring home for the British Museum.

With two American trips behind him, on 4 February 1848, Lyell filled London's Royal Institution for one of its noted Friday evening lectures. An audience of 400, including most of the geologists in the capital and many literary figures as well, turned out to hear him speak on 'The fossil reptile footprints in Pennsylvania'.

Lyell told how he had been shown the reptile prints at Greensburg, Pennsylvania, a town about twenty miles east of Pittsburgh, where they lay in the coal formation of the Allegheny Mountains. He had been taken out to a stone quarry by a young doctor, Dr Alfred King, along with a local Presbyterian clergyman. (Mrs Lyell had remained at the inn in Greensburg where, he noted, the landlady was especially kind to her.) When Dr King had first made the fossil discovery, he was fiercely denounced by a Catholic clergyman for denying the account of creation in the Bible. However, the Protestant clergyman who accompanied him, and saw the prints with his own eyes, kept silent.[41]

Lyell was astounded. He could see at once that the footprints were genuine. Apart from being those of a quadruped, they recalled the bird tracks he had just seen in Nova Scotia. He concluded that the tracks had been produced as the reptile walked over the clay-like mud before the mud had begun to dry and crack.

Before his London audience, Lyell displayed one of the slabs he had brought back with him. He pointed out the differences between the toe-grouping in the prints and those of the European *Cheirotherium* (the 'hand beast' whose tracks resembling human hands had first been found in Germany in 1833). Each pair of the Pennsylvanian

Sir Henry De la Beche.

Duria Antiquior, by Henry De la Beche. Watercolour painted in 1830 based on fossils found in Lyme Regis by Mary Anning.

Amgueddfa Genedlaethol Cymru – National Museum of Wales

Mary Anning and her dog, Tray, painted before 1842.

The skull of *Temnodontosaurus platyodon*, an ichthyosaurus that lived between 201 and 194 million years ago in the Lower Jurassic. It was discovered by Mary Anning's brother in 1811 in Lyme Regis.

Adam Sedgwick, 1867.

Gideon Mantell, c. 1850.

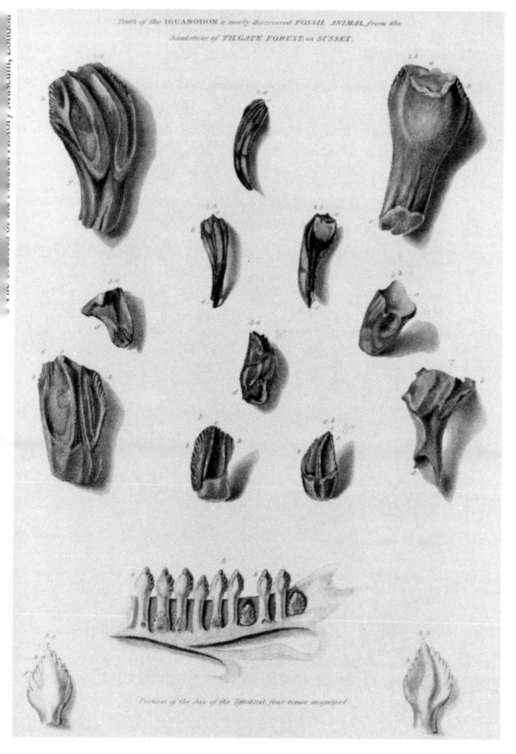

Original drawings by Gideon Mantell showing the iguanodon fossil teeth that he discovered. These drawings were published in the *Philosophical Transactions of the Royal Society*, pl 14, 1825.

Sir Roderick Impey Murchison, 'Men of the Day No. 14', published in *Vanity Fair*, 26 November 1870.

Ogigiocarella, an Ordovician trilobite.

Louis Agassiz.

Charles Darwin as a young man. Watercolour by George Richmond, painted in 1840.

footprints suggested a creature that had walked with a protuberance like another toe.

For his Royal Institution listeners, with the Duke of Northumberland in the chair, Lyell stressed the importance of these footprints. While coal strata had often been seen to hold remains of plants, no traces of any air-breathing, non-marine creatures had ever before been detected in rocks so old. He stressed that it was safe to assume that the huge reptile which left those prints on the ancient sands of the coal measures was an air-breather, for 'its weight would not have been sufficient under water to have made impressions so deep and distinct'.[42] The message he was bringing was one that the geological world was ready to learn: life had developed differently on the opposite sides of the ocean; North America had had its own kind of quadrupedal ancestor to the dinosaur.

In conclusion, Lyell made a political as well as a scientific point. He summarised the dilemma that the Pennsylvania footprints presented for scientific discourse. In the United States, with its free press, religious toleration and social equality, he felt that there was still insufficient intellectual freedom to enable a student of nature to discuss with impunity the philosophical questions presented by science. He called for the British to set up 'a good system of primary schools' to make impossible 'that collision of opinion, so much to be deprecated, between the multitude and the learned'. In sum, he gave a rousing call for mass education and a free press.

Lyell was in his best form. His publisher John Murray had just paid him £250 for the seventh edition of *Principles* (1847). His *Travels* (1845) and *A Second Visit to the United States* (1849) make for interesting reading. He continued to expound on his fascination with coalfields and unusual fossils, always interlacing his observations with a compulsive reporting of American society, its immigrants, population rise and mix of races. He testified to the great future resources of the country. He would make two more trips to the United States before his attention was required by Darwin and Prince Albert, for more pressing projects in his own country.

15

AT LAST, THE BIG QUESTION

In June 1846 Lyell returned from his second trip to North America certain of the superiority of secular universities. His own country, he told American friends, was 'more parson-ridden than any in Europe except Spain'.[1]

He had ended his first volume of colourful American travel stories with an attack on Oxford and Cambridge. Why, he asked, 'should we crowd all the British youth into two ancient seats of learning? Why not promote the growth of other institutions in London, Durham, Scotland, Wales, and Ireland?'[2] It would be desirable for Oxford and Cambridge to expand freely and cease to be places for educating the clergy of the Established Church. In his eyes, the two venerable universities as they stood encouraged 'class division and exclusive sectarianism'.

In February 1847 twenty-seven-year-old Prince Albert was elected chancellor of Cambridge University and set himself to broaden the curriculum. The prince consort, an ardent supporter of science, had been reading Lyell's *Travels*. When Lyell sent him the seventh edition of *Principles*, he was rewarded with an invitation to Buckingham Palace.

Lyell, then in his early fifties, had met the young prince a few years earlier and now found his English much improved.[3] Meeting on 28 March 1847, they chatted animatedly about a cause dear to them both: university reform. Indeed, classical languages and mathematics were still held to be the only subjects suitable for Cambridge or Oxford. Neither university had woken up to the new industrial need for trained scientists. For Lyell and the prince, their agreement on this important issue was the beginning of a good relationship. On a visit to the palace the following year, Lyell told Albert how he admired American public (or state) education. He feared that many people in England opposed schooling for the lower classes as it might make them discontented with their lot. He told the prince that in America, slave owners used the same argument for the same reason.[4]

The two men had more to discuss than university reform. Albert wanted Lyell to sit on the royal commission for the Great Exhibition, which was still in its planning stages. The country weathered the Chartist agitation of April 1848, when a crowd of more than 100,000 (some accounts say 300,000) gathered on Kennington Common in south London demanding universal suffrage. Long-held fears of a duplication of the revolutions sweeping Europe led Queen Victoria to flee London for her safety. A guard was mounted at scientific institutions, among them the Geological Survey where Henry De la Beche brought in an armful of cutlasses and prepared for a siege.[5] The Reverend William Buckland waited at Westminster Abbey armed with a crowbar. However, the mass protest meeting was no uprising and the throng dispersed peacefully.

In the late summer of 1848, Queen Victoria summoned Lyell to Balmoral to receive a knighthood. (Lyell would become the third member of the Geological Society to be knighted. Humphry Davy in 1812 had been the first; the second – not until 1846 – was Roderick Murchison.) Prince Albert wanted Lyell to be the first person knighted at Balmoral, the castle in Aberdeenshire on which he had bought a long-term lease from its owner, Lord Aberdeen. For Lyell, it was a straight fifty miles north on horseback across the Grampian Hills from his father's Kinnordy estate. He was invited to stay overnight, although not at Balmoral itself; accommodation was arranged at an inn about a mile away, with a carriage sent by the queen to take him back and forth. Together Lyell and the prince, as Lyell wrote to his sister, 'had a most agreeable geological exploring on the banks of the Dee'.[6]

When what was called 'The Great Exhibition of the Works of Industry of all Nations' opened in Hyde Park on 1 May 1851, Prince Albert took justifiable pride in the triumphant realisation of his bold vision. He called it a 'living picture of the point of development at which the whole of mankind has arrived'.[7] The massive and unprecedented glass structure of 'the Crystal Palace', designed by Sir Joseph Paxton, was an astounding success. There was a glittering opening by Queen Victoria. The exhibits celebrated, as works of art, the machinery, scientific instruments and technical inventions of the industrial revolution. Before the exhibition's closure five months later, on 15 October, some six million people had attended, including Darwin, Lyell and George Eliot. The cheapest tickets were a shilling. Public toilets could be visited for one penny (a charge that gave rise to the prim and enduring euphemism, 'to spend a penny').

The exhibition earned £186,000 above its costs (over £16 million today). Its success led Albert to establish the subsequent 'Surplus Committee', on which Lyell served, to decide what to do with the profits. One objective of the committee was to set up a trust to provide grants and scholarships for industrial research; another was to develop a nearby tract of land. The museums of South Kensington, in an area now sometimes referred to as 'Albertopolis', stand as a monument to the princely vision.

In the early summer of 1850, John Murray's publishing house was preparing to bring out a new edition of Lyell's *Elements of Geology* (this 1838 book being distinct from Lyell's earlier and most popular work, *Principles of Geology*). Lyell went to Down House in Downe, Kent, to discuss with Charles Darwin what changes might be required in *Elements*. Lyell was now serving his second two-year term as president of the Geological Society. A mark of Lyell's acknowledgement of his debt to the eighteenth-century James Hutton was his choice of cover for the new edition: a woodcut of

Scotland's Siccar Point, where Hutton had first glimpsed the spectacularly mismatched rocks that confirmed his belief in the vast age of the earth.

In January 1852 the fourth edition of yet another popular work by Lyell, *A Manual of Elementary Geology*, appeared – its sales speeded by the success of his American travel books. For this edition Lyell wrote a new introduction in which he raised for the first time the big question: why and how did new species appear?

In March the Lyells took themselves to Hyde Park to take a last look at the now empty glass-and-iron Great Exhibition Hall. Mary Lyell found it beautiful: 'so very large there is something quite dreamy about the extent of it & the light colouring'.[8] They also attended the queen's ball at Buckingham Palace, where they watched Victoria and Prince Albert dancing in a quadrille, 'enjoying themselves very much'.

That summer saw a smallpox outbreak in London; Mary made sure that everyone in their household at Harley Street was vaccinated. In July they celebrated their twentieth wedding anniversary and met Charles Dickens for tea. Weeks later Lyell made a speech at the reopening of the Crystal Palace, which had been rebuilt in Sydenham, where it stood until fire destroyed it in 1936. (Later, as a government commissioner reporting on the 1851 exhibition, Lyell would reiterate his message about the need for continuing the tradition of large industrial exhibitions. He claimed that what had been seen in London had 'created a unity of all the nations of the world, however different their tendencies and systems of government'.) After spending the weekend with the Darwins, they left in August for their third trip to America – their first as Sir Charles and Lady Lyell.

On 6 October 1852, Sir Charles went with James Hall, a New York geologist, to western Massachusetts to examine the erratic boulders in the Berkshire Hills. The stones were extraordinarily large – one was 52 feet long and 40 feet wide – standing in long parallel straight rows. The usual explanation – glacial action – would not do. Lyell decided that the erratics could not have been carried by glaciers because the boulders were distributed at right angles to the ridges above them. In this he was wrong; the boulders had been carried by continental glaciers rather than by the mountain glaciers Lyell knew about. (He was also struck by the New England farmhouses made of wood – something unknown in Britain – and by their porches holding rocking chairs.[9])

Delivering his third series of Lowell lectures in Boston in November, Lyell acknowledged the geological sophistication of his audience, who knew very well that the periods of which he was speaking extended back for millions of years. For the first time in his lectures Lyell addressed the species question. Why, he asked them to consider, had certain species disappeared? Fossils showed that the North American mastodon had lived in quite recent geological times. Why had it vanished? Then he raised the tougher question: how to account for the appearance of new forms of life? Mixing the theological with the geological in the dozen Lowell lectures of this series he examined theories for the emergence of new species. Did they replace those gone extinct? He offered three possibilities: first, the Lamarckian idea of 'transmutation' – emergence through changes in pre-existing species; second, new species simply appearing unrelated to others; third (his own favoured idea), they were brought into being by a creative act of God.

In the very first edition of *Principles*, Lyell had provided a chapter called 'The Progressive Development of Life'; but the chapter was long and tortuously argued – single paragraphs extended over two pages – and his otherwise elegant prose was befuddled by the inability to say anything clear and outright. But from this earliest work, Lyell had insisted upon the recent origin of man. He did not accept that there were any human bones among the fossils of extinct mammals found in cave deposits. In his view, claims that some of these were human remains were unauthenticated. To him, man was an entirely different species. As he had summed up in his postscript to the fourth edition of *A Manual of Elementary Geology*: 'Physically considered, he [man] may form part of an indefinite series of terrestrial changes past, present, and to come: but morally and intellectually he may belong to another system of things – of things immaterial – a system which is not permitted to interrupt or disturb the course of the material world, or the laws which govern its changes.'[10]

Lyell titled his twelfth and final Lowell lecture 'Progressive Development'. He accepted that one class of organisms might replace another, but held that there was no evidence of preparation of the earth to receive man nor – his old theme – of any grand catastrophe. He ended with more praise to the 'rational lord of Creation'.

In 1853, in the ninth edition of *Principles*, Lyell gingerly expanded his discussion of the disappearance of species. As a man of religious faith, he had shied away from the mystery of mysteries: how had man emerged? He pleaded a religious excuse: 'To assume that the evidence of the beginning or end of so vast a scheme lies within the reach of our philosophical inquiries, or even of our speculations, appears to us inconsistent with a just estimate of the relations which subsist between the finite powers of man and the attributes of an Infinite and Eternal Being.'[11]

Yet what laws governed their emergence? Perhaps each year one new species appeared and an old one became extinct?[12] He speculated, tentatively edging into the big question of: 'whether the human species is one of the most recent of the whole?'[13]

It was in September the following year, 1854, that Charles Darwin recorded in his journal: 'Began sorting notes for Species Theory'. Two more years would pass before he revealed to Lyell, his closest friend, that he had his own ideas on the emergence of new species: that they appeared through the process of natural selection. Darwin's proposals directly contradicted Lyell's own beliefs, but as a scientist of utmost integrity, Lyell accepted that, in science, existing ideas must give way when superseded by new discoveries.

16

ORIGIN OF *ORIGIN*

Clergymen who engaged in the fashionable hobby of fossil-collecting also had to consider why certain species had vanished and, more unsettlingly, whether the new were more advanced than the old. In 1840 the Reverend George Young, a Scottish Presbyterian clergyman and scholar who had discovered an ichthyosaur in 1819, posed the delicate question in 'Scriptural Geology' – an essay on the high antiquity ascribed to organic remains found imbedded in stratified rocks. 'Some have alleged,' wrote Young, 'that in tracing the beds upwards we discern among the enclosed bodies a gradual progress from the more rude and simple creatures, to the more perfect and completely organised; as if the Creator's skill had improved by practice.' But he immediately backed away, saying: 'But for this strange idea there is no foundation: creatures of the most perfect organisation occur in the lower beds as well as the higher.'[1]

The idea that man had evolved from lower forms of life was so controversial that perhaps the first book boldly putting it in print was published anonymously. In 1844 *Vestiges of the Natural History of Creation* offered the proposition that fossils showed life had advanced through a series of stages by natural means. 'The simplest and most primitive type . . . gave birth to the type next to it and so on to the very highest.'[2] *Vestiges*, later known to have been written by the Scottish publisher and editor, Robert Chambers, was an intelligent and lengthy summary of geological history. It concluded that, considering the whole system of nature, 'we cannot well doubt that we are in the hands of One who is both able and willing to do us the most entire justice'.[3]

Vestiges, with what one critic called its 'clear, pleasant, racy, self-sufficient style', rode the wave of scientific popularisation started by Lyell's *Principles* and also by the Scottish geologist Hugh Miller's *Old Red Sandstone* in 1841. Edition after edition of *Vestiges* was quickly snapped up by an eager public. While generally well received by Victorian society, the book infuriated the clergy. In the *Edinburgh Review*, the Reverend Adam Sedgwick exploded in rage and rhetoric. In an anonymous essay of which he was widely known to be the author, he filled over eighty-five pages to attack *Vestiges* for telling people that they were not made in the image of God but were the children of apes. 'Degrading materialism', he called the book. He was particularly worried about its appeal to young women and wrote to a friend that 'God willing, I will strive to abate the evil.'[4] Sedgwick seemed to forget that he himself had told a Geological Society audience in an anniversary address in 1834 that in 'the repeated and almost entire changes of organic types in the successive formations of the earth . . . we have a series of proofs that there has been a progressive development of organic structure subservient to the purpose of life'.[5]

Vestiges galvanised the new Young Men's Christian Association, formed the year the work was published. The YMCA began a series of free public lectures intended to protect young men from the threats to faith posed by the book, especially after it became more widely available in a cheap format. The religious newspaper *Christian Observer* let fly a volley of insults: the book was 'undisguised materialism', 'atheistic in its tendencies', even 'pigology'.[6] The questions the book raised went beyond the physiological. If humans were descended from monkeys, where did their sense of morality come from? How could they be expected to exercise independent judgement?

The mystery of its authorship was ended in 1884 with the twelfth edition of *Vestiges*. Published after Chambers's death, it carried his name on the cover, spine and title page.

The question of progress had dogged geology from the beginning. Were the older rocks without fossils inferior to the fossiliferous laid down later? Speculation inexorably shifted to life itself. Were tiny trilobites a step on the way up to dinosaurs? Was humankind the destination of the changes shown in fossils? How did one species lead to another – if 'lead' it did? What had caused the mass extinction of the dinosaurs 65 million years ago? And did mammals then grow in size because there was no competition for eating the vegetation?[7]

The wider general understanding was expressed in Alfred Tennyson's *In Memoriam A.H.H.*. Completed in 1849, the poem is thought to have been inspired by *Vestiges of Creation*. Written to express Tennyson's unhealing grief for his friend Arthur Henry Hallam, who had died of a brain haemorrhage in 1833, the poem succinctly summarises the new awareness of the vastness of time and the indifference of nature to man. Man might trust that:

> . . . God was love indeed
> And love Creation's final law–
> Tho' Nature, red in tooth and claw
> With ravine shriek'd against his creed–

but he knew also that human existence and love were nothing to an impersonal, relentless nature which allows forms of life to disappear:

> 'So careful of the type?' but no.
> From scarped cliff and quarried stone
> She cries, 'A thousand types are gone:
> I care for nothing, all shall go . . .'[8]

The pessimism of the poem did not trouble Queen Victoria. After the sudden death of Prince Albert in 1861 at the age of forty-two, she found great consolation in Tennyson's grieving conclusion: 'Tis better to have loved and lost / Than never to have loved at all.'

The queen took some comfort also in a private visit for which she had summoned Sedgwick in 1863. As he had known her and Albert in their happiness at Osborne, she asked him to come to see her at Windsor Castle. He believed he was the first person outside her own family to whom she fully opened her heart and told her sorrows. As he wrote in his journal: 'After the first greeting, when I bent one knee and kissed her hand, there was an end of all form, and the dear sorrowing Royal lady talked with me as if I had been her elder brother,' – Sedgwick was then seventy-eight. 'Her great aim is to carry out the intentions of the great and good Prince whom God has removed from her side. "He had the greatest regard for you," she said, "and that is why I had a strong desire to talk with you without reserve."'[9]

Darwin traced his own first thoughts on the origin of species to the late 1830s, well before the publication of *Vestiges*. In November 1839 he wrote to Henslow, his mentor in Cambridge, that he was 'steadily collecting every sort of fact, which may throw light on the origin & variation of species'.[10] By 1842 he had written the first sketch of his idea that new species were produced by natural selection. Two years later he wrote out a publishable essay on the subject in case he died before completing the full version, which he wanted to be on the scale of Lyell's *Principles*.

His work on evolution may possibly have sprung from his poor health. Had he been in full strength and vigour, Darwin might have continued as a geologist hammering rocks in the open air rather than plunging into speculative theorising in his study. His appalling ailments, which included palpitations and chronic vomiting, prevented fieldwork and turned him into a reclusive country gentleman. His attention turned to breeding flowers (orchids in particular) in his quiet Kent garden and greenhouse and also to describing and classifying barnacles from all over the world. As always he was thorough. In 1854 he informed Sir Joseph Hooker that he was 'sending ten thousand Barnacles out of the house all over the world . . . I shall in a day or two begin to look over my old notes on species'.[11] The bespectacled Hooker, whom Darwin told Lyell was 'a most engaging young man', would become Darwin's close friend.[12] As deputy director of the Royal Botanic Gardens at Kew he knew very well that Darwin's enthusiasm for gardening and horticulture was shared by thousands of middle-class Victorians. Indeed, the whole country seemed gripped by a passion for greenhouses and exotic plants.

The temptation to read progress into the fossil record had long existed – Lamarck having called attention in 1809 to the manner in which the characteristics of species changed over generations. For religious believers, in any event, the meaning of the fossil record was always obvious: a succession of constant changes led to the divinely ordered destination: man.

An eccentric solution to the apparent contradiction between fossils and God's handiwork was offered in 1857 by Philip Henry Gosse, a field naturalist and respected anatomist who spent much time in Devon. Two years before Darwin's great book, Gosse came up with the theory that God had created the world to look as if it had already existed, placing fossils in the rocks as evidence of a 'past', just as He had created Adam with a navel even though the First Man had neither umbilical cord nor mother. Gosse's book was called *Omphalos* (Greek for 'navel') from its central idea, with its arch subtitle: *An Attempt to Untie the Geological Knot*. Fossils, in Gosse's reasoning, were not evidence of a past but rather of God's intention to make the world appear old. (When *Omphalos* came out, it was ridiculed. Yet some Jehovah's Witnesses today subscribe to this theory of the deceptive age of fossils.) Gosse, needless to say, had no patience with *Vestiges of the Natural History of Creation* and its assumption that the fossil record showed the world's long history. Of the anonymous author, he joked 'this writer has hatched a scheme, by which the immediate ancestor of Adam was a Chimpanzee, and his remote ancestor a Maggot!'[13]

On 18 June 1858, the following year, Darwin received a shock from the other side of the world: a devastating letter from another naturalist, Alfred Russel Wallace. Darwin instantly saw that Wallace, a much-travelled explorer who had felt out of place in London and was now working in Borneo, had independently come up with the theory of natural selection through the struggle for existence.

There can be no greater measure of Darwin's integrity than what he did in response. Rather than throwing the letter away, as he could easily have done, he complied with Wallace's request and passed the essay on to Lyell. (Wallace knew Lyell only from his *Principles*.) In his covering letter to Lyell, his closest friend, Darwin wrote: 'Your words have come true with a vengeance that I shd. be forestalled. You said this when I explained to you here very briefly my views of "Natural Selection" depending on the struggle for existence.' Lyell had urged Darwin to present a simple sketch of his views, but Darwin declined, saying that to give a fair sketch would be impossible without supporting his propositions with facts. He admitted, 'yet I should certainly be vexed if any one were to publish my doctrines before me'.[14] 'Though,' Darwin added, 'my Book, if it will ever have any value, will not be deteriorated, as all the labour consists in the application of the theory.'[15]

Lyell could see that the radical new idea of human evolution had emerged through his own description in *Principles* of the gradual accumulation of changes through minor adaptations. He urged Darwin to hurry up and get on with writing and publishing what his friend now called his 'big book on species'. Indeed, the unfinished book was already 250,000 words long. After a polite letter to Wallace explaining that he himself was writing a large book on species and variation, Darwin got back to work.

Haste was in the air. Two weeks later at a meeting of the Linnean Society of London on 1 July 1858, Darwin's theories on natural selection were presented by the secretaries of the society, preceded by introductory notes prepared by Lyell and Darwin's good friend Hooker. The Linnean Society was the chosen venue because the subject of both papers – living species – was considered unsuitable for the rock-centred Geological Society.

At the meeting, the introductory comments made clear that Darwin had already expressed his ideas on divergence of species in an essay sketch sent in a letter to a Harvard botany professor, Asa Gray. These ideas were now being laid out fully in the big work Darwin had in progress. The thesis of this new book was then explained, followed by a reading of the essay Wallace had sent to Darwin. The sequence of the Linnean programme made Wallace's work seem a 'me-too' offering, subordinate to Darwin's original thinking. Darwin was, as usual, not present at the meeting. Charles, his tenth and last child, had died of scarlet fever, increasing his (justified) anxiety that his family, by marrying their close cousins as many of them had done, had inflicted genetic weakness on their children.

The Darwin and Wallace papers were first on a long programme at the Linnean Society. Read aloud, one after the other, they passed without discussion. With too much to digest, the members and guests left exhausted. The news that he and Darwin had been given simultaneous airing was posted to Wallace, who was surprised to learn that his own original idea had also occurred to Darwin. From then on, both men – undisputedly co-discoverers of one of nature's great truths – behaved with exceptional courtesy and tolerance.

Both papers were published in the Linnean Society journal in September 1858. Asa Gray brought the ideas to Louis Agassiz at Harvard. Agassiz was surprised to learn that Gray was inclined to believe the theory, because he himself was passionately opposed to the very idea of evolution. When later he read Darwin's book, he pronounced it 'poor – very poor!!'

Darwin got down to writing in earnest and finished the 500-page manuscript in May 1859. As he worked he tightened his style. According to his biographer Janet Browne, he omitted all footnotes, compressed his material, discovered a voice that was 'in turn dazzling, persuasive, friendly, humble, and dark', and produced a masterpiece of readable, genial scientific text.[16] For the process he was describing, he did not use the word 'evolution' but rather 'descent with modification'. It was his friend, Herbert Spencer, who in 1864 gave Darwin's theory the label, taken from Darwin's text, which has stuck: 'survival of the fittest'. Darwin's actual words were: 'we must suppose that there is a power, represented by natural selection or the survival of the fittest, always intently watching each slight alteration in the transparent layers; and carefully preserving each which, under varied circumstances, in any way or in any degree, tends to produce a distincter image'.[17]

The ground had been prepared by *Vestiges of the Natural History of Creation*. But Chambers's book made the development of fossils appear to lead towards something higher on the scale of life, while Darwin carefully omitted any sense of progress or purpose or striving towards a goal. Unlike Chambers, he filled his book with scientific facts and did not discuss how life began.

Darwin was perhaps more worried about how his own life could continue. Two months before publication, his vomiting started again and he took himself off for a water cure or, as he called it, 'hydropathy and rest'.

When Darwin asked Lyell's advice on bringing out his new book, Lyell steered him towards his own publisher and fellow Scot, John Murray. Murray undoubtedly knew that controversy was good for business. Very soon Darwin's butler was on his way to Murray's offices in Albemarle Street, bearing a brown-paper parcel containing the battered manuscript.

Lyell gave his friend Darwin's book a powerful blast of advance publicity. Speaking in Aberdeen in September 1859 (after lunching with Queen Victoria and Prince Albert at Balmoral), he delivered a public lecture to the British Association for the Advancement of Science, alerting his scientific audience to the imminence of a work to be published in two months' time which would throw 'a flood of light on many classes of phenomena connected with the affinities, geographical distribution and geological succession of organic beings, for which no other hypothesis has been able, or has even attempted to account'.[18]

The British Association speech was a public acknowledgement that Lyell would stand by Darwin through the storm he saw about to break. In his lecture he spelled out the important new evidence uncovered by excavations in France: Stone Age tools that indicated human beings were coexistent with fossils of mammals which lived about 2.5 million years ago.

On the Origin of Species by Means of Natural Selection, or the Preservation of Favoured Races in the Struggle for Life hit the world on Thursday 24 November, 1859. George Eliot and her partner George Henry Lewes read it on the Saturday. The book opened with a tribute from Darwin to Lyell: 'He who can read Sir Charles Lyell's grand work on the *Principles of Geology* . . . yet does not admit how incomprehensibly vast have been the past periods of time, may at once close this volume.' Eliot's own tribute came the following year with the publication of her novel *The Mill on the Floss*, whose concluding scene features a great flood sweeping away Maggie and Tom, sister and brother, reconciling all their past differences as they drown in an embrace.

Darwin's work is one of the greatest books of all time. In clear language it deals with the sudden appearance of species after the long Primary period represented by the earlier rocks from which fossils were absent. It describes what is now understood as 'the Cambrian explosion' (545 million years ago), when a great burst of varieties of life, trilobites not least but also other forms of arthropods (invertebrate animals having an external skeleton, a segmented body and jointed appendages), left their traces in the rocks. Darwin wrote simply: 'the amount of organic change in the fossils of consecutive formations probably serves as a fair measure of the relative, though not actual lapse of time'.[19] Then, as 'all the living forms of life are the lineal descendants of those who lived before the Cambrian epoch, we may feel certain that the ordinary succession by generation has never once been broken, and that no cataclysm has desolated the whole world'.[20]

There followed perhaps his most famous sentence (now engraved on a wall of the Royal Society's dining room): 'It is interesting to contemplate a tangled bank, clothed with many plants of many kinds, with birds singing on the bushes, with various insects flitting about, and with worms crawling through the damp earth, and to reflect that these elaborately constructed forms, so different from each other, and dependent upon each other in so complex a manner, have all been produced by laws acting around us.'[21]

Thus the struggle for life was 'a consequence to Natural Selection, entailing Divergence of Character and the Extinction of less improved forms'. Then came his peroration: 'There is grandeur in this view of life, with its several powers, having been originally breathed by the Creator into a few forms or into one; and that whilst this planet has gone cycling on according to the fixed law of gravity, from so simple a beginning endless forms most beautiful and most wonderful have been, and are being evolved.'[22]

The book proved the bestseller Murray's had hoped. It aroused international interest and went into many editions – six by 1876. The first review, using the words that Darwin had not, claimed that he had said that men came from monkeys.

As usual, Adam Sedgwick came up with the fiercest critical blast. Not exhausted by condemning *Vestiges*, he attacked his old student (for that is how he regarded Darwin) after the publication of *Origin*. He wrote to Darwin himself to tell him that he had read the book 'with more pain than pleasure': 'Parts of it I admired greatly, parts I laughed at till my sides were almost sore.' But other parts he read 'with absolute sorrow, because I think them utterly false and grievously mischievous – You have deserted – after a start in that tram-road of all solid physical truth – the true method of induction – &

started up as wild, I think, as Bishop Wilkin's locomotive that was to sail us to the moon.' He felt that Darwin's argument rested on 'assumptions which can neither be proved nor disproved'. Darwin used 'natural selection' as if it implied 'causation' but omitted to explain the cause. For himself, Sedgwick said, 'I call (in the abstract) causation the will of God; and I can prove that He acts for the good of His creatures. He also acts by laws which we can study and comprehend.' But to him, Darwin had ignored the link between the material and moral. In essence, what was the cause of the development of species of which he wrote?[23]

He also disliked Darwin's tone of triumphant confidence – the same tone he had deplored in *Vestiges of Creation*. Darwin's reasoning would 'sink the human race into a lower grade of degradation than any into which it has fallen since its written records tell us of its history'. What Darwin had to do was to accept God's revelation.[24]

Signing off his letter as 'a son of a monkey and an old friend of yours', Sedgwick reported that his health was somewhat better but that he 'humbly accepted God's revelation of Himself both in His works and in His word . . . If you and I do all this, we shall meet in heaven.'[25]

Speaking to others, Sedgwick was fiercer, calling Darwin's book 'a dish of rank materialism cleverly cooked and served up'.[26] He called Darwin himself 'the teacher of error instead of the apostle of truth'. The book ignored the 'God of Nature' as manifested in His works and denied the moral and metaphysical aspects of humanity. It was an unsurprising response from a clergyman who every Sunday attended three services and preached twice.

Wallace, for his own part, in a long letter to Darwin sent from the East Indies warmly praised the book of which Darwin had sent him a copy and generously wrote:

> As to the theory of Natural Selection itself, I shall always maintain it to be yours and yours only. You had worked it out in details I had never thought of, years before I had a ray of light on the

> subject, and my paper would never have convinced anybody, or been noticed as more than ingenious speculation, whereas your book has revolutionised the study of Natural History and carried away captive the best men of the present age. All the merit I claim is having been the means of inducing you to write and publish at once.[27]

The success of *On the Origin of Species* did not take the sting out of Darwin's recognition in 1861 that he had been completely wrong about Scotland's Parallel Roads of Glen Roy. Long before, in 1847, Darwin had written to Hooker: 'I have been bad enough for these few last days, having had to think & write too much about Glen Roy (an audacious son of dog (Mr Milne) having attacked my theory) which made me horribly sick.'[28]

What David Milne, a Scottish geologist, had done was to visit Lochaber, believing Darwin's sea-beaches theory to be correct. Yet when he got to Glen Roy he discovered a gap previously unnoticed on the level of the middle road. He gave the gap the unpoetic name of 'Col R2' ('col', from the Latin '*collum*' for neck, a geological term for a pass or gap between two mountain peaks). Milne (later Milne-Home) recognised the col, correctly, as the gap through which the overflowing glacial water had escaped from the lake.

What Darwin and his predecessors (notably the Scottish geologists John MacCulloch, FRS, and Sir Thomas Dick-Lauder, whose 1818 paper 'The Parallel Roads of Glenroy' first drew attention to Scotland's geological mystery) lacked was the essential information: knowledge of glaciers and glaciology – the origins of the roads being finally explained by the palaeontologist Agassiz in 1840.

In 1892, looking back at the prolonged search for the Glen Roy explanation, John Tyndall commented at the Royal Institution on the manner in which the origins of the roads had been considered by Darwin and other eminent geologists of an earlier time: 'Two distinct mental processes are involved in the treatment of such a question. Firstly, the faithful and sufficient observation of the data; and secondly, that higher mental process in which the constructive imagination comes into play, connecting the separate facts of observation with their common cause, and weaving them into an organic whole.'[29] Tyndall called attention to the handicap under which Darwin and the early Glen Roy investigators had laboured: 'A knowledge of the action of ancient glaciers was the necessary antecedent to the next explanation, and experience of this nature was not possessed.'[30]

And the vital lesson? According to Tyndall: 'In the survey of such a field two things are especially worthy to be taken into account – the widening of the intellectual horizon and the reaction of expanding knowledge upon the intellectual organ itself.'[31]

The rise of aggregated scientific thinking, characterised by the theories of the Glen Roy roads from 1776 to 1876, shifted the balance between speculation and observation. The Parallel Roads showed what could be achieved by collective thought built on known causes and rational deduction: observation, deduction, supposition and calculation were now the ways to approach a problem. A purely Romantic view was a hindrance.

The rise of the scientific study of landscape in the early nineteenth century coincided with the demise of Romanticism. The relationship between science and literature showed the difference between the instructed and the uninstructed imagination. Romantic literature teemed with impulse, spirit and will. In contrast, science was interested in origins, outcomes and predictions. In essence, literature gradually withdrew from the material world throughout the nineteenth century. While William Wordsworth and Samuel Taylor Coleridge were looking inwards, Darwin, Lyell and other scientists looked outwards – hence their visits to Glen Roy.

But, unlike poets, scientists could be proved wrong. Darwin had to accept the evidence on Glen Roy. But not until 1861 did he finally surrender: 'I give up the ghost. My paper was one gigantic blunder.'[32]

But he never had to use the word 'blunder' about *Origin of Species.*

17

THE WHOLE ORANG

Did Lyell realise, after the publication of *Origin*, that his pupil was now his teacher? Without acknowledging that they had changed places, with intellectual honesty Lyell swiftly and completely rewrote his *Manual of Elementary Geology* for its tenth and much-enlarged edition in 1867, spelling out the evidence of evolutionary ancestry for man.

Evolution remained controversial. Five years after *Origin*'s publication, in November 1864 – when, after some of the older members were afraid, in Lyell's words, 'of crowning anything so unorthodox as the "Origin"'[1] – the Royal Society's distinguished Copley Medal was awarded to Darwin. (Darwin as ever did not attend.) The citation cagily praised Darwin's achievements in geology, physical geography, zoology, physiological botany and genetic biology, but deliberately omitted any mention of *On the Origin of Species*. Evolutionary theory was still unmentionable in polite scientific society.

Lyell gave the after-dinner speech usually given by the medallist. Darwin wrote and thanked him. In reply Lyell related to Darwin what he had told the Royal Society: 'I said I had been forced to give up my old faith without thoroughly seeing my way to a new one. But I think you would have been satisfied with the length I went.'[2]

Lyell was a reluctant evolutionist. He believed in a creator and the distinctiveness of man, who was 'of higher dignity than were any pre-existing beings on the earth'. Yet his attention had been caught by the discovery of flint tools that seemed to him evidence of primitive man. In 1859 he had taken himself to France to inspect the gravel beds of the Somme and Abbeville as well as to question palaeontologists who knew the site, with its indications that humans had lived in the same era as mammoths and cave bears. The flint tools discovered among fossils, he said, were some signs of 'the likely barbarism of early humanity'.[3] He found the tools contemporaneous with the mammoth and wrote to his friend George Ticknor that he regarded 'the Pyramids as things of yesterday in comparison of these relics'.[4] He obtained sixty-five of the tools and was pleased when a former governor of New Zealand recognised them as similar to spearheads found in Australia and hatchets such as the natives of Papua New Guinea used for digging up roots.[5]

Four years later, Lyell elaborated on the theme in *The Antiquity of Man*. In this detailed book subtitled *The Geological Evidences of the Antiquity of Man, with Remarks on Theories of the Origin of Species by Variation*, he traced the fossil evidence of living creatures from the earliest Cambrian era to the present day, and recorded in one work all the documented evidence on human antiquity back to the eighteenth century. He considered his latest book to be the story of the relations of the early history of man to the glacial period – the period crucial to understanding the origin of man.[6] In April 1863 the Royal Society's *Proceedings* supported his argument with its report on the first human jawbone ever discovered, found in the same French gravel pit that contained the ancient flint tools.

Published by Murray, *Antiquity* sold out almost immediately in the first week of February 1863 and was reprinted three times that year. Mudie's Circulating Library bought several thousand copies for a public hungry to read about the controversial process of evolution. According to the Darwin biographer Janet Browne, *Antiquity* was 'the first significant book after Darwin's *Origin* to shake humanity's view of itself'.[7] It gave people their geological past. Darwin, however, poring over its pages, could not find any open endorsement of evolution.

From 53 Harley Street, on 11 March 1863, Lyell wrote to Darwin to explain his position. He knew that the public 'have regarded me as the advocate of the other side' – that is, anti-evolution and transmutation – but, he told Darwin, 'you much overrate my influence': 'My feelings, however, more than any thought about policy or expediency, prevent me from dogmatising as to the descent of man from the brutes . . . I cannot admit that my leap at p. 505 [in *Antiquity of Man*] which makes you "groan" is more than a legitimate deduction . . . I think the old "creation" is almost as much required as ever, but of course it takes a new form if Lamarck's views improved by yours are adopted.'[8]

Lyell remained convinced that there was a huge gulf between man and beast. How the gulf had been bridged remained 'a profound mystery'.[9]

'Oh,' commented Darwin in the margin of his copy.[10]

In *Antiquity*, Lyell avoided any mention of God by name. Instead he employed graceful allusions to 'the Author of Nature', 'the Supreme Cause' or 'the Supreme Will and Power'. He had been unable to abandon his religious beliefs, telling Darwin in May 1869: 'I feel that progressive development or evolution cannot be entirely explained by natural selection, I rather hail Wallace's suggestion that there may be a Supreme Will and Power which may not abdicate its functions of interference, but may guide the forces and laws of Nature'.[11] He concluded *Antiquity* with a ringing endorsement of what is now called 'intelligent design'. Evolution, he wrote, 'leaves the argument in favour of design, and therefore of a designer, as valid as ever'. He reminded his readers that Isaac Newton, originator of the theory of gravitation, was religious as well as philosophical: 'the improvable reason of Man himself presents us with a picture of the ever-increasing dominion of mind over matter'.[12]

Darwin invited the Lyells for the weekend at Down House to celebrate the publication of Lyell's book, but he wrote to his confidant Joseph Hooker that he was 'deeply disappointed at Lyell's excessive caution': 'the best of the joke is that he thinks he has acted with the courage of a Martyr of old.—I hope I may have taken an exaggerated view of his timidity . . . I wish to Heaven he had said not a word on the subject.'[13] Yet when his health forced cancellation of the visit, Darwin wrote a candid letter to Lyell saying Lyell was not doing enough to help him.

Lyell, smarting at the charge, wrote his own answer to Hooker: 'Darwin has sent me a useful set of corrigenda and criticisms for the new edition I am busy in preparing. He seems much disappointed that I do not go farther with him, or do not speak out more. I can only say that I have spoken out to the full extent of my present convictions, and even beyond my state of feeling as to man's unbroken descent from the brutes, and I find I am half converting not a few who were in arms against Darwin and are even now against Huxley.'[14]

The question would haunt Lyell until his death. He could not bring himself to go 'the whole orang', as he sardonically put it to the biologist and writer T. H. Huxley.

Huxley in 1868 coined the word 'agnostic' to describe his own position as neither believer nor disbeliever. He had become known as 'Darwin's bulldog' for his passionate defence of Darwin's theory of evolution by natural selection. Darwin had shared his theory with him before publication, positioning Huxley to become the leading supporter of the theory.

In a celebrated confrontation at Oxford at the new Gothic-style Museum of Natural History, where the British Association for the Advancement of Science was meeting on 30 June 1860, the great public speaker Samuel Wilberforce, Bishop of Oxford, ridiculed Darwin's theory. Wilberforce had earlier attacked Robert Chambers's *Vestiges*. On this occasion he asked Huxley (accounts of the famous exchange vary, usually relying on Hooker's account in a long letter to Darwin) whether he would prefer a monkey for his grandfather or his grandmother.[15] Huxley's retort has variously portrayed him as calm, as white with anger, or simply calling Wilberforce 'unscientific'. In Lyell's account Huxley replied 'that if he had his choice of an ancestor, whether it should be an ape, or one who having received a scholastic education, should use his logic to mislead an untutored public, and should treat not with argument but with ridicule the facts and reasoning adduced in support of a grave and serious philosophical question, he would not hesitate for a moment to prefer the ape'.[16] However, he would be ashamed to be connected with a man who used his great gifts for 'the mere purpose of introducing ridicule into a grave scientific discussion'. (The British Association quickly moved to provide a sanitised account of the afternoon.[17])

Another surprising Darwin critic that day was the former captain of the *Beagle*, Robert FitzRoy, now head of the government's Meteorological Department. As Hooker described it, 'lifting an immense Bible first with both hands and afterwards with one hand over his head, [FitzRoy] solemnly implored the audience to believe in God rather than man'. FitzRoy then admitted that *Origin of Species* had given him 'acutest pain'.[18] He had already written to Darwin – 'My dear old friend' – that 'I, at least, cannot find anything "ennobling" in the thought of being a descendant of even the most ancient Ape.'

Hooker sent Darwin an exuberant report of the confrontation – which had been so heated that one woman fainted and crowds waited outside the door hoping to get in. However, the meeting appears to have ended cheerfully, with all going off to dine together afterwards. Each speaker thought he had won. Yet the confrontation lived on in history as the defining moment when religion faced the claim of science to explain the origin of human beings.[19] But the question remained open: were humans descended from monkeys or were they created by God?

For Lyell, it was not so much the hope of an afterlife to which he clung but rather to his belief that man was unique – not just another creature on the biological chain. He did not hold this belief out of any sentimentality as a father. The contrast between him and Darwin in fecundity was glaring. There is no record that Lyell ever alluded to his childlessness; one wonders if he and Darwin, with their mutual interest in reproduction, ever discussed it. Darwin seems to have taken to heart Thomas Malthus's stricture that contraception was a vice. 'Moral restraint' was the only permissible barrier to pregnancy.[20] Charles and Emma Darwin did not practise much restraint. By the time *Origin* was published, they had seven living children, three others having died.

Despite their differences, Darwin retained his respect for his mentor. So he should have done. When the tenth edition of Lyell's *Principles*, much enlarged, came out in two volumes in 1867 it revealed that Lyell had indeed changed his view. After what he called 'Mr. Darwin's epoch-making work', he wrote that he now accepted the descent of man from ape.[21]

In the year of Lyell's death, his friend Hooker paid a moving tribute to his intellectual courage: Lyell had abandoned 'a theory, which he had for forty years regarded as one of the foundation-stones of a work that had given him the highest position attainable among contemporary scientific writers . . . and to substitute a new foundation not only more secure, but more harmonious in its proportions than it was before'.[22]

The world of science was completely converted to Darwin's views. In 1869 the publisher Alexander Macmillan founded a weekly scientific journal, *Nature*, with a polemical purpose: to argue Darwin's scheme. Half a dozen articles in *Nature*'s first year did just that, and Darwin became a lifetime subscriber.[23]

The friendship between Darwin and Lyell endured as Lyell's health failed and Mary Lyell died in the spring of 1873. In November 1874 when Lyell attended the fiftieth anniversary of the Geological

Society Club, of which he had been a founding member, his friends were surprised at his vigour. He died not long after, on 22 February 1875, at the age of seventy-eight. By then blind, he had fallen down the stairs at his Harley Street home.

Darwin lamented the loss of the best friend he had ever had. He was glad that he had died with his faculties intact and praised his 'freedom from all religious bigotry . . . How grand, also, was his candour & pure love of truth.'[24] He gave his valedictory words on Lyell to his friend Hooker (now president of the Royal Society): 'How completely he revolutionised Geology: for I can remember something of pre-Lyellian days. I never forget that almost everything which I have done in science I owe to the study of his great works.'[25] Yet when Hooker arranged for the great geologist to have the honour of burial in Westminster Abbey, Darwin declined the invitation to be a pallbearer. He could not even bring himself to go to the funeral – for fear of becoming ill, he said. As he put it: 'I should so likely fail in the midst of the ceremony, and have my head whirling off my shoulders.'[26] Thus Lyell was buried with benefit of clergy, but without that of Darwin.

In its commemoration, the journal *Nature* praised Lyell's decision to give up law for geology: 'it was well for science that he was induced to prefer the quieter study of nature to the subtle bandying of words or the excitement of forensic life'.[27] Lyell knew very well what he had done. As he had said in a letter to Thomas Spedding in May 1863, 'the question of the origin of species gave much to think of, and you may well believe that it cost me a struggle to renounce my old creed'.[28]

But renounce it he had.

18

MUSEUM PIECES

Charles Lyell's legacy lies in his writings. His *Principles of Geology* and *Travels in North America* retain a freshness and intelligence today. Lyell's significance was acknowledged at Westminster Abbey on his tombstone (which was made of Carboniferous limestone and small fossils). For the inscription Sir Joseph negotiated with Lyell's sister Katherine a form of words that was not too religious: 'Throughout a long and laborious life he sought the means of deciphering the fragmentary records of the earth's history in the patient investigation of the present order of Nature enlarging the boundaries of knowledge and leaving on scientific thought an enduring influence. "O Lord how great are thy works and thy thoughts are very deep". Psalm XCII.5.'[1]

The abbey holds Darwin's final resting place as well. When Darwin died in February 1882, he was buried in the nave – not as close to Lyell as his wife Emma had wished, but appropriately near those other giants of British science, Sir John Herschel and Sir Isaac Newton. Among the pallbearers were his old friends Hooker, Huxley and Wallace. At the time, Darwin's stone was inscribed only with his name and dates. Later the Royal Society added words on his contribution to science. Huxley led the memorial committee to pay homage to the distinguished scientist and £4,500 was raised, half of which was used for the statue in the Natural History Museum in South Kensington. Darwin's white marble form now sits like a deity at the top of the main staircase at the museum. For years, it looked down at the school children crowded around the symbol of the museum's collection – the giant skeleton of the dinosaur Diplodocus – although in 2016 this was taken down, to be replaced by a blue whale, in a decision to focus on the living, not the fossil past.

When after eight years' construction the great Victorian museum, designed by Alfred Waterhouse and created by the driving force of Richard Owen, opened in April 1881, *The Times* called it the 'Temple of Nature, showing, as it should, the Beauty of Holiness'.[2] It had been established as a separate entity from the British Museum when the natural history collections became so large as to require a home of their own. In a new twenty-first-century wing, the museum offers the less beautiful but technically advanced Darwin Centre, which allows visitors to see into the museum's laboratories and to interact with exhibits.

On the ground floor is a sight not to be missed. Displayed along a gallery wall is the *Plesiosaurus giganticus* that Mary Anning dug out of a cliff in 1823 – along with her brother Joseph's great discovery of 1811, the wide-eyed skull of the first ichthyosaur, and also her own 1830 find, the *Plesiosaurus macrocephalus*. On the wall hangs a portrait of Anning in a black cloak and straw bonnet, with her dog Tray; she is described as 'the greatest fossil-hunter ever known'.[3] The museum also offers a costumed 'Mary Anning', who talks about helping to discover the first specimen of an ichthyosaur when she was only ten.

Those who visit Down House in Kent are rewarded with a strong sense of where and how the reclusive Darwin lived for forty years with Emma and his children. The modest Victorian villa, built in the 1730s but later modernised, is an English Heritage site and one of the main tourist attractions in southeast England. It holds his library, his study with his chair, notebooks, microscopes, and other scientific instruments, and his greenhouse where he raised plants to study their adaptations. There is also a replica of his cabin on the *Beagle*. The garden holds the paths where he walked, especially during the ailing last years of his life.

The man to whom Darwin was apprenticed in his youth, the Reverend Adam Sedgwick, died in Cambridge in January 1873 at the age of eighty-seven, having finished his Norwich residence only months before and returned happily to his beloved Trinity College, Cambridge. He died with a prayer on his lips, heard by his niece: 'Wash me clean in the blood of the Lamb – Enable me to submit to Thy Holy Will – Sanctify me with Thy Holy Spirit.'[4]

Sedgwick remained as Woodwardian Professor of Geology to the last. As early as 1850 he had written: 'after thirty laborious geological tours, I have brought together and placed in the Cambridge Museum, a very noble Collection'.[5] Indeed he had, placing his acquisitions in the Woodwardian Museum. Three years before his death he expanded his intentions: 'There were three prominent hopes which possessed my heart in the earliest years of my Professorship,' he wrote. 'First, that I might be enabled to bring together a Collection worthy of the University, and illustrative of all the departments of the Science it was my duty to study and to teach. Secondly, that a Geological Museum might be built by the University, amply capable of containing its future Collections; and lastly, that I might bring together a Class of Students who would listen to my teaching, support me by their sympathy, and help me by the labour of their hands.'[6]

Upon Sedgwick's death, the university decided that his memorial should take the form of a new and larger museum. The Sedgwick Museum of Earth Sciences was opened in 1904, with King Edward VII in attendance. Now part of the university's Department of Earth Sciences, it offers a superb collection of fossils, rocks and minerals from around the world.

For the last part of his life, the Reverend William Buckland lived in London as Dean of Westminster. He dropped out of active geological research, yet still encouraged a respect for the natural sciences within the Church of England. He continued to give the annual lecture course (begun in 1814) on geology at Oxford until 1849, when mental disability forced him to retire. He then moved to Islip, northeast of Oxford. In the churchyard, in what was perhaps his last joke, he reserved a plot which he knew lay over an impenetrable outcrop of solid Jurassic limestone. Upon his death in 1856, a hole had to be blasted to accommodate his wishes and his coffin.

The Buckland Collection forms the centrepiece of the Oxford University Museum of Natural History's holdings of zoological, entomological and geological specimens. His prized *Megalosaurus* jaw is on display there. Among the other treasures of the collection is a cast of the ferocious marine predator *Mosasaurus*, bearing an inscription: 'Given by the Museum of Natural History at Paris to Dr Buckland'. The donor was identified as Georges Cuvier of the French Musée Nationale, who made the gifts as evidence of his friendship and the high esteem in which he held Buckland.

Oxford also holds (to the dismay of the National Museum of Wales) the remnants of Buckland's celebrated find, inescapably if inaccurately known as 'the Red Lady of Paviland'. Although analysis long ago revealed the bones to be those of a male, genetic facts have not dimmed the Red Lady's fame. The bones are carefully stored in the museum's basement.

Cardiff nonetheless does hold one of the most important geological archives in the world. Henry De la Beche in 1837 moved his Geological Survey to Swansea because of the economic importance of the Welsh coalfields. Knighted in 1842, he became active in the local scientific scene; his daughter Elizabeth married the son of a Swansea naturalist, Lewis Weston Dillwyn (of the family who led Buckland to the Paviland Cave in Gower). Dillwyn later became mayor of Swansea.

De la Beche died in April 1855 aged fifty-nine, having spent his last years in a wheelchair suffering from progressive paralysis. He is buried in Kensal Green Cemetery in London. More than a century later documents revealed that an illegitimate daughter, Rosalie Torrie, is buried beside him. On De la Beche's death his early papers, journals and sketches remained with the Dillwyn family, who in the 1930s gave them to the National Museum of Wales. The vast collection includes some of De la Beche's finest art works, notably the brilliant *Duria Antiquior*, the watercolour created for Anning in 1830. The museum also holds significant Buckland papers. Given through the same Dillwyn family connections, they show details of his discovery of the Paviland Cave.

In Cambridge, Massachusetts, Harvard's Museum of Comparative Zoology, founded by the efforts of Louis Agassiz in 1859, is stuffed with prehistoric and extinct animals, many on display, including the largest turtle shell ever found, a fifty-foot-long skeleton of a mastodon and another of a killer whale. The museum, sometimes called the Agassiz Museum, also houses contributions from a dozen academic departments reflecting the new breadth of zoology, from biological oceanography to vertebrate palaeontology.

With his great map, William Smith made his own memorial in 1815. One of the original copies hangs reverentially, like the 'Last Supper' of geology, in the entrance hall of the Geological Society's headquarters in Burlington House, London, with curtains drawn to protect it from the light. The map is not small: covering such a wide area of Britain, emphasising sequence of strata, and drawn on a scale of five miles to the inch, it reaches more than eight feet in height and over six feet across.[7] There is also a fine portrait of Smith, painted in 1837, two years before his death at the age of seventy.

As for Roderick Murchison, despite all the geographic features around the world bearing his name (including the Murchison Oil Platform in the North Sea), the only Murchison museum appears to be in the town of Murchison on the west coast of South Island in New Zealand, where it was built to commemorate the earthquake that devastated the area in 1929. A large slab of rock bearing the inscription: 'To Roderick Impey Murchison, Scottish geologist, explorer of Perm Krai, who gives to the last period of Palaeozoic era the name of Perm' stands in the central Russian city of Perm. Murchison is more grandly remembered in London, with a grave in the Brompton Cemetery covered by an imposing multi-layered engraved tomb.

19

THEN AND NOW

One could weep for all that the Victorian geologists did not know that is common knowledge today. Darwin, who died in 1882, never heard of genes or chromosomes, nor did other scientists of his time.

But science is a continuous process of discovery. Darwin could never know the secret of life found in 1953 by James Watson and Francis Crick at the Cavendish Laboratory in Cambridge. They recognised that genetic information is transferred from parent cells to new through the twisted and paired threads of the nucleic acid, DNA. More cause for regret is that Darwin and his scientific contemporaries worked without the benefit of a paper published in 1866, now considered the foundation of modern genetics. In *Experiments with Plant Hybrids*, Gregor Mendel, an Austrian monk, described the transmission of hereditary characteristics, basing his observations on the results of crossing thousands of pea plants. Yet Mendel's paper appeared in an obscure journal and was largely ignored at the time. Indeed, not until the end of the nineteenth century were the terms 'gene' and 'genetics' coined by William Bateson, a Cambridge student of heredity who recognised the strength of Mendel's ideas and brought them to scientific attention.

What is also now understood is why Lamarck was wrong. Changes in structure did not arise from new conditions – despite the dogged insistence today by Rupert Sheldrake, biologist and researcher into parapsychology, that the giraffe's long neck was produced by countless generations stretching their necks to reach for fresh foliage on higher branches and the mole's blindness by burrowing in the earth and not needing the use of eyes. The reason why some organisms survived over others is that the fittest survived in the competition for food, not that individual organs adapted themselves to life's conditions.

Another major discovery was that of the Burgess Shale in 1909 in the Canadian Rockies. It was found by a young palaeontologist, Charles Doolittle Walcott, and is considered the most spectacular fossil deposit anywhere in the world.[1] This rich fossil bed first opened the world's eyes to the vast array of animals that had been living in the Cambrian seas 505 million years ago. The profusion in the area of trilobites and soft-bodied animals with no hard parts was found during the building of a railway hotel in Yoho National Park and was described as 'the finest and largest series of Middle Cambrian fossils yet discovered in any formation in any country'.[2] Walcott, a self-made man who, without a degree in geology, became director of the United States Geological Survey and secretary of the Smithsonian Institution in Washington, DC, is now considered to have contributed more than anyone before or since to the understanding of the life of the Cambrian world.

The condescension of hindsight quickly vanishes in the face of the certainty that today's scientific knowledge will be superseded even more quickly than yesterday's. Almost every weekly issue of the still-thriving *Nature* carries new information that renders some previous knowledge obsolete. 'Two-billion-year-old fossils are the first sign of multicellular life,' the journal asserted on 1 July 2010. Fossils large enough to be seen with the naked eye were found in southeastern Gabon in rocks known to be 2.1 billion years old. 'Early Life', a report in *Nature* by P. C. J. Donoghue and J. B. Antcliffe, concluded that: 'The discovery and continuing elucidation of the Precambrian fossil record has met Darwin's predictions on the extent and structure of evolutionary history.'[3]

From the beginning of geology, calculating the age of the earth has been a preoccupation, and the Comte de Buffon's guess at 75,000 years, derived from his experiment with heating and cooling iron spheres, had earned its place in the science's history. By 1870, however, *Nature* could declare that it had 'become possible that, by means of changes which are known to have occurred in a given number of years, some measurement of the time represented by the whole series of geological formations might be obtained'. These changes had occurred very slowly, the journal conceded, 'yet not so slowly as to be quite imperceptible in historical time'.[4]

The actual age of the earth was determined in the mid-twentieth century by the dating of meteorites formed at the same time as the planet. The distinguished geologist Arthur Holmes (1890–1965) of Edinburgh University and fellow of the Royal Society and Imperial College found a way to use the radioactive decay of uranium to measure the age of rocks. He had performed the first radiometric dating while on the staff of Imperial College, the result being published in 1911. He used the fact that radioactive decay converts an isotope of uranium into a form of lead at a predictable rate. He calculated that 'a gramme-molecule of lead would take the place of a gramme-molecule of uranium in 8,200 million years'.[5] With this decay rate established, scientists had for the first time a clock by which to tick off the millennia that had passed since the elements had been formed. According to this clock, which was less well calibrated at the time, he first estimated the age of the planet as 3 billion years in 1927, and this has been corrected to an estimate of 4.54 billion years today.

Yet not until 2.45 billion years ago did oxygen become a tiny component of the earth's atmosphere, and built up to modern levels much later. The cause of what is now dubbed 'The Great Oxidation Event' is unknown but the event is believed to have originated from algae which poured oxygen into the atmosphere, making possible the appearance of more complex, air-breathing forms of life.

The age of the human race is proving harder to calculate. The dates for the appearance of the Neanderthals and others who shared common ancestors with modern man keep changing; evidence has emerged that they used stone tools about 800,000 years earlier than had been thought. The discovery of sharp-edged shaped bones in Ethiopia has suggested that early people chipped stones to make them more effective in cutting and scraping bits of flesh from the carcasses of animals. Our ancestors can thus be seen to have been butchers and carnivores about 3.2 million years ago.

As for the universe itself, the search for its age continues. In 1929 the American astronomer Edwin Hubble worked out that the galaxies are flying apart as a consequence of the Big Bang with which the universe originated. Allan Sandage of the Carnegie Observatories in California, who worked with Hubble, corrected Hubble's calculations and first estimated that the age of the universe was 15 billion years, which was then recalculated to be 13.7 billion years. What is beyond doubt is that the universe is expanding.

One thing had not changed as geology entered its third century. Geology versus Genesis still rages, despite all the territory apparently secured by the former. 'Scriptural Geology', the Reverend Young's essay of 1840, has been republished as a paperback and is available on bookselling websites such as Amazon. In 2008 in Florida, the State Board of Education found it necessary to rule (by the narrow margin of four to three) that 'while belief in a Divine Creator of the universe is a religious belief, the scientific theory that higher forms of life evolved from lower ones is not'. Sarah Palin, the former Republican US vice-presidential candidate, argued for teaching 'intelligent design' as science in schools. A British group called 'Truth in Science' works to introduce 'creationism' and 'intelligent design' into schools as a counterbalance to the requirement of the National Curriculum for England, Wales and Northern Ireland, that students be taught that 'the fossil record is evidence for evolution'. At the same time Britain's leading scientific polemicist, Richard Dawkins, declared in his 2006 bestseller, *The God Delusion*, that teaching creationism to children is a kind of child abuse. A solution (one that displeased Dawkins) was suggested in an editorial in *Nature* that asked what scientists might do to counter the appeal of 'intelligent design' (as distinct from the more literal 'creationism'): 'All scientists whose classes are faced with such concerns should familiarise themselves with some basic arguments as to why evolution, cosmology and geology are not competing with religion. When they walk into the lecture hall, they should be prepared to talk about what science can and cannot do, and how it fits in with different religious beliefs.'[6] Dawkins and many others – not just scientists – feel that is needlessly conciliatory, claiming too little for the astonishing insights of modern science and offering an apology for the endlessly changing nature of scientific knowledge it establishes where none is needed.

For all the efforts of the original geologists to distinguish the Noachian deluge from natural processes, some still believe that the Bible's account of the Flood is literally true. A new speciality called 'Flood Geology' is flourishing, particularly among creationists in the United States. Its believers have searched (so far in vain) for remnants of Noah's Ark on Mount Ararat.[7]

What has been shown is that in ancient times there was indeed a massive flood – yet not in the geographical place the Bible suggests, but rather in what is now the Black Sea. In 1999 *Noah's Flood: The New Scientific Discoveries about the Event that Changed History*, by William Ryan and Walter Pitman, claimed that 7,000 years ago a land barrier gave way and the Mediterranean Sea rushed through the Dardanelles, severing Asia from Europe, turning a pre-existing lake into the Black Sea. Archaeologists had previously assumed that the historic reality behind cultural spread of the deluge legend had been an exceptional flood of the Tigris and Euphrates rivers. But many are now persuaded that 'the Flood' was in fact the far more ancient drowning of the pre-Black Sea lake, a traumatic natural disaster that would explain the cultural spread of the Flood legend.

Another mystery closer to solution is the extinction of the dinosaurs. The Cretaceous period, beginning about 145 million years ago, was their heyday. The earth was much warmer than it is today and it was a period of great geological and biological unrest. And it ended, according to *Nature*, 'in spectacular style, with the global catastrophe that saw off dinosaurs some 65 million years ago, an event known as the Cretaceous/Palaeogene (K/Pg) extinction'.[8] All the evidence shows that the devastation was caused by an asteroid striking the Mexican Yucatan Peninsula and sending up a vast dust cloud that blacked out the sun and caused long-term freezing and lack of light. The event has been redated to 66 million years ago.

Of even wider popular interest is the unending debate about whether the complexity of the universe – or, indeed, of the multiplicity of universes which seem to keep appearing – is evidence of God as creator. Religious faith in an age of science seems more than ever to be a matter of personal belief rather than of persuasion by the complexity of the cosmos. In 2010 Stephen Hawking all but apologised for having ended his bestselling *A Brief History of Time* in 1988 with the mischievous assertion that if we knew the theory of everything which explained the existence of the universe 'then we should know the mind of God'.[9] The famous line may have helped sales, he concedes, but in *The Grand Design* (written in 2010 with Leonard Mlodinow), he maintains that modern physics leaves no room for God. The equally famous and openly atheistic scientist Richard Dawkins has never wavered in his unbelief – a view shared by many scientists.

The biggest change in understanding the planet earth has come with the discovery of plate tectonics. In 1915 the German geologist Alfred Wegener proposed the theory of continental drift: that pieces of the earth's crust move slowly over a fluid mantle. He maintained that the planet originally held only one giant continent, to which he gave the name 'Pangaea' (Greek for 'whole earth'), and that Pangaea started to break into two smaller continents – he called them 'Laurasia' and 'Gondwanaland' – 250 million years ago.

Wegener was fascinated by the glaring fact that on a map the Atlantic Ocean looks as if its two sides had been pulled apart. To him, the configuration suggested that the Atlantic was an enormously widened rift with the sides still matching 'as closely as the lines of a torn drawing would correspond if the pieces were placed in juxtaposition'.[10]

Arthur Holmes, commenting on Wegener's ideas, wrote on the closing pages of his classic *Principles of Physical Geology*, first published in 1944: 'The parallelism of the opposing shores of the Atlantic has been a subject of discussion ever since Francis Bacon first drew attention to it in 1620.'[11] In papers in the 1920s and 1930s he had mentioned continental drift as the cause but the idea was not accepted until the 1960s. American geologists were particularly reluctant to accept it.[12] Holmes's influential textbook, however, endorsed continental drift in its last summary chapter.

In the 1960s the 'widening Atlantic' idea became refined and deepened as the theory of plate tectonics. It is now beyond question that the various plates of the earth's crust drift, collide, slip past or under each other. This resolved the long controversy over how mountain belts were thrown up, whether by sudden change or inch by inch. Under the seas, the plates spread from mid-oceanic rift zones where magma from within the earth's mantle comes to the surface. When one plate slips under the other, intensive folding forms high mountain ridges (like the Andes and the Himalayas) and magma can rise and form volcanoes.

It is now accepted that there were supercontinents before Pangaea and that others will follow. In his 2007 book *Supercontinent*, the geoscientist Ted Nield has written: 'the world we see today is no more than Pangaea's smashed remains, the fragments of the dinner plate that dropped on the floor'.[13]

Geologists have found evidence that between 635 and 716 million years ago sea ice extended to the equator – bolstering the theory that at times the planet has been covered with ice at all latitudes and that some of these Ice Ages lasted for millions of years. '"Snowball Earth" Confirmed: Ice Covered Equator', the *National Geographic* declared in 2010.[14] The discovery was made by Harvard University scientists who, studying volcanic rocks in Canada, found these rocks sandwiched between glacial rocks known from previous magnetic studies to have formed when Canada was near the Equator. Thus, over time – a great length of time – the movement of the earth's tectonic plates pushed the rocks to northwest Canada.

One thing is certain: the permanence of change. It has been calculated, says Nield in *Supercontinent*, that in another 200 million years or so, a new supercontinent will be born.[15] Well before that, Africa will get a new ocean. In northern Ethiopia, the Great Rift Valley (one of the hottest places on earth) is splitting apart. The rift is sixty kilometres long, and, according to the summer exhibition of the Royal Society in 2010, marks the birthplace of Africa's new ocean.

Over time, the sides of the Atlantic will continue to grow farther apart. This spread occurs along the mid-Atlantic ridge, stretching along the ocean floor from the Arctic to the Antarctic. In the middle of this ridge is a north–south deep valley, or rift.

A pleasant place to observe this phenomenon is Iceland. The mid-Atlantic ridge runs down the middle of the island and lies above sea level. It includes a deep rift, into which visitors, standing on a rocky promontory high above Gullfoss, 'the golden waterfall', can stare down and marvel, some getting vertigo from the clarity of the water. Before their eyes, although scarcely perceptible, the North American and the Eurasian plates are pulling apart at the rate at which fingernails grow.

Trying to understand the planet's rocks and oceans is a challenge that will endure. New knowledge will constantly wipe out the certainties of the past. Geologist and writer Richard Fortey has wryly observed that the first-century Roman historian Suetonius 'would have dismissed me as a madman had I pointed out to him that Vesuvius is a consequence, ultimately, of Africa moving bodily northwards'.

But Africa really is on the move. As Fortey explains: 'The northern edge of the African plate is plunging beneath the southern tip of Italy which twists and deforms the rocks and uplifts the limestone hills.' Volcanoes then rise up 'like sticky blood oozing from deep wounds where the earth's crust thinned'. Everything around Naples, Fortey says, is controlled by the movements and interactions of tectonic plates far below the surface.[16]

What would Charles Lyell have made of that? A sense of exhilaration, of revelation, of vindication?

At the very least, a new edition of *Principles*, no doubt.

NOTES

FOREWORD

1 Brenda Maddox, *George Eliot: Novelist, Lover, Wife*, London: Harper Press, 2009; *George Eliot in Love*, New York: Palgrave Macmillan, 2010.
2 Cited in Janet Browne, *Charles Darwin*, Vol. 2: *The Power of Place*, New York: Alfred Knopf, 2002, p. 63.
3 Martin J. S. Rudwick, *The Great Devonian Controversy*, Chicago: University of Chicago Press, 1985, p. 37.

CHAPTER 1 THE ABYSS OF TIME

1 Charles Lyell, *Principles of Geology*, edited with an introduction by James A. Secord, London: Penguin Classics, 1997, p. 6 (referred to hereafter as Lyell/Secord, *Principles*).
2 Katherine Murray Lyell (ed.), *Life, Letters and Journals of Sir Charles Lyell, Bart.*, London: John Murray, 1881, Vol. 2, p. 14.
3 Roy Porter, 'Gentlemen and Geology: the Emergence of a Scientific Career, 1660–1920', *The Historical Journal*, 21.4, Dec 1978, p. 815; Ralph O'Connor, *The Earth on Show: Fossils and the Poetics of Popular Science, 1802–1856*, London and Chicago: University of Chicago Press, 2007, p. 193.
4 James Secord, citing *Spectator*, 14 Jan 1832, introduction, Lyell/Secord, *Principles*, p. 39.
5 Richard Holmes, *The Age of Wonder: How the Romantic Generation Discovered the Beauty and Terror of Science*, London: Harper Press, 2008, p. 244.
6 Holmes, *Age of Wonder*, p. 289.
7 Jack Repcheck, *The Man Who Found Time: James Hutton and the Discovery of the Earth's Antiquity*, Cambridge, Mass., Perseus, 2003, pp. 17–18.
8 Cited in Holmes, *Age of Wonder*, p. 296.
9 See Victor R. Baker, 'Catastrophism and uniformitarianism', London: Geological Society, Special Publications 1998, v. 143, p. 174.

10 Leonard G. Wilson, *Charles Lyell: The Years to 1841: The Revolution in Geology*, New Haven and London: Yale University Press, 1972, p. 138.
11 Lyell to Mantell, 3 Nov 1831, cited in Christopher McGowan, *The Dragon Seekers: How an Extraordinary Circle of Fossilists Discovered the Dinosaurs and Paved the Way for Darwin*, Cambridge, Mass.: Perseus 2001, p. 44.
12 Ibid.
13 Lyell to Mantell, Mar 1831, Lyell (ed.), *Life, Letters and Journals*, Vol. 1, p. 317.
14 Lyell/Secord, *Principles*, p. xxiv, citing Lyell to Scrope, 14 Jun 1830, in Lyell (ed.), *Life, Letters and Journals*, p. 268.
15 Lyell to C. Baggage, 3 May 1832, in Wilson, *Revolution in Geology*, p. 353.
16 Lyell to Mary Horner, 1 May 1832, in Lyell (ed.), *Life, Letters and Journals*, Vol. 1, p. 353.
17 Ibid., p. 354.
18 Lyell to Mary Horner, 31 Dec 1831, in ibid., p. 360.
19 Lyell to Mary Horner, 17 Feb 1832, in ibid., p. 344.
20 Adrian Desmond and James Moore, *Darwin's Sacred Cause: Race, Slavery and the Quest for Human Origins*, London: Allen Lane, 2009, p. 161.
21 Lyell/Secord, *Principles*, p. 30.
22 Bill Bryson, *A Short History of Nearly Everything*, London: Black Swan, 2003, pp. 103–4, observes that many cite the date as 26 October 4004 BC. Others differ in the timing, placing it either on the evening of 22 October, or midday on 23 October.
23 Jim Endersby, 'Creative Designs', *Times Literary Supplement*, 16 Mar 2007.
24 Dennis R. Dean, *James Hutton and the History of Geology*, Ithaca, New York: Cornell University Press, 1992, p. 29. In the introduction to Lyell/Secord's *Principles* (p. 452, n. 7) Secord describes Lyell's 'celebrated misquotation' – 'In the economy of the world, I can find no traces of a beginning, no prospect of an end' – as 'the most celebrated statement in the history of the earth sciences'. He attributes it to Lyell's using Playfair as a text.
25 Hutton cited in Lyell/Secord, *Principles*, p. 14.
26 Dean, *James Hutton*, p. 29.
27 Lyell/Secord, *Principles*, p. 15; and Repcheck, *The Man Who Found Time*, p. 65.
28 Lyell/Secord, *Principles*, p. 234.
29 Review of Buffon by Jacques Roger in *Nature*, 6 Nov 1997, Vol. 390, p. 37.

30 Richard Fortey, 'Lyell and Deep Time', *Geoscientist*, 9 October 2011, Vol. 21, No. 9, p. 14.
31 James Hutton, *Theory of the Earth with Proofs and Illustrations*, Vol. 1, Edinburgh: William Creech, 1795, p. 183.
32 See Charles Coulton Gillispie, *Genesis and Geology: The Impact of Scientific Discoveries upon Religious Beliefs in the Decades Before Darwin*, New York: Harper & Row, 1951; HarperTorchbook, 1959.
33 Ibid. It may be noted that the tempting combination of the two words still flows easily off the tongue; books such as *Genesis and Geology* continue to appear.
34 John Playfair, *The Works of John Playfair, with a memoir of the Author*, 4 vols, Edinburgh: Archibald Constable and Co., 1822, p. 80.
35 Stephen Jay Gould quoted in Prologue of Jack Repcheck, *The Man Who Found Time*, p. 1.
36 Stephen Jay Gould, *Time's Arrow, Time's Cycle: Myth and Metaphor in the Discovery of Geological Time*, Cambridge, Mass.: Harvard University Press, 1987, p. 2.
37 Dr Iain Stewart, 'Men of Rock' documentary, BBC4, broadcast 6 Jan 2011.

CHAPTER 2 HEALTHFUL EXERTION

1 George Henry Lewes, *Sea-Side Studies,* Edinburgh: William Blackwood & Sons, 1858, p. 11.
2 Cited in C. L. E. Lewis and S. J. Knell (eds), *The Making of the Geological Society of London*, London: Geological Society, 2009, p. 14.
3 Robert Bakewell, *An Introduction to Geology*, London: J. Harding, 1813, p. 22.
4 Ibid., p. 361.
5 John F. W. Herschel, *Preliminary Discourse on the Study of Natural Philosophy*, London, 1831, p. 287.
6 Lyell/Secord, *Principles*, pp. 24–5.
7 Ibid., p. 25.
8 Porter, 'Gentlemen and Geology', p. 850.
9 Cited in Richard Holmes, *Coleridge: Early Visions*, London: Hodder and Stoughton, 1989, pp. 66–7.
10 John Davy (ed.), *The Collected Works of Sir Humphry Davy*, Vol. 1, London: 1839, p. 50.
11 William Wordsworth, *The Excursion*, London: Edward Moxon, 1836, p. 83.

12 Lyell/Secord, *Principles*, p. 10.
13 Ibid., p. 11. The spelling 'Freyberg' is Lyell's own.
14 Honoré de Balzac, *The Wild Ass's Skin*, Everyman Paperback, 1967, pp. 19–29.
15 Holmes, *Coleridge: Early Visions*, p. 353.
16 John Davy (ed.), *The Collected Works of Sir Humphry Davy*, Vol. 1, London: 1839, p. 466.
17 Sedgwick to his American cousins, Mrs Norton and Miss Sedgwick, 8 Jul 1869, in J. W. Clark and T. M. Hughes (eds), *The Life and Letters of the Reverend Adam Sedgwick*, London: C. J. Clay and Sons for Cambridge University Press, 1890, Vol. 1, p. 123.

CHAPTER 3 DOWN THE MINES

1 George Orwell, *The Road to Wigan Pier*, Chapter 2. The actual quote reads 'Our civilization, pace Chesterton, is founded on coal . . .'
2 Lyell/Secord, *Principles*, p. 9.
3 W. S. Jevons, *The Coal Question*, London: Macmillan, 1866, p. 376.
4 Matt Ridley, 'Want Cheap Energy?', *The Times*, 5 May 2011.

CHAPTER 4 VESTIGES OF PATERNITY

1 Simon Winchester, *The Map That Changed the World*, London: Penguin, 2002, pp. 91–2.
2 A. Sedgwick, 'Address to the Geological Society', *Proceedings of the Geological Society of London*, London: the Geological Society, 1831, Vol. 1, p. 368.
3 Clark and Hughes (eds), *Life and Letters of the Reverend Adam Sedgwick*, Vol. 1, p. 368.
4 Gordon L. Herries Davies, *Whatever is Under the Earth: the Geological Society of London 1807–2007*, London: Geological Society, 2007, p. 79.
5 Winchester, *Map That Changed the World*, p. 68.
6 Ibid., p. 81.
7 Ibid., p. 2.
8 Ibid., p. 41n.
9 Ibid., p. 179n.
10 Deborah Cadbury, *The Dinosaur Hunters*, London: Fourth Estate, 2000, p. 94, citing Horace Woodward, *The History of the Geological Society of London*, London: Geological Society, 1907.
11 Lyell/Secord, *Principles*, p. 22.
12 Ibid., p. 22, quoting J. F. D'Aubuisson des Voisins, *Traité de géognosie*, Vol. 2, p. 253.

13 Cuvier, 1812, '*L'histoire de ce monde*', in O'Connor, *The Earth on Show*, p. 62.
14 Gideon Mantell to Benjamin Silliman, 14 Jun 1841 in Leonard G. Wilson, *Lyell in America: Transatlantic Geology*, 1841–53, New Haven and London: Yale University Press, 1998.
15 Cited in Stephen Jay Gould, *Dinosaur in a Haystack: Reflections on Natural History*, London: Cape, 1996, p. 165.
16 Wilson, *Revolution in Geology*, pp. 116–17.
17 Lyell (ed.), *Life, Letters and Journals*, Vol. 1, p. 185.
18 Lyell/Secord, *Principles*, p. 78.
19 Lyell, review of *Memoir on the Geology of Central France*, by G. P. Scrope, London 1827, *Quarterly Review*, 1827m, Vol. 36, pp. 437–83.
20 Lyell to his sister Marianne, 20 Oct 1828, Lyell (ed.), *Life, Letters and Journals*, Vol. 1, p. 209.
21 Lyell to his sister Eleanor, 9 Nov 1828, ibid., p. 213.
22 Ibid.
23 Cited in Wilson, *Revolution in Geology*, p. 253.
24 Lyell to Murchison, 12 Jan 1829, in Lyell (ed.), *Life, Letters and Journals*, Vol. 1, p. 234.
25 Lyell to Murchison, 15 Jan 1829, in ibid. p. 234.
26 Lyell/Secord, *Principles*, p. 152.
27 The image he labelled a temple is now thought to have been a marketplace. However, it served an almost religious purpose for Lyell.
28 Lyell/Secord, *Principles*, p. 70.
29 Ibid., p. 149.
30 For an example of this well-known observation, see Walter Gratzer, *The Undergrowth of Science: Delusion, Self-Deception, and Human Frailty*, Oxford: Oxford University Press, 2000.
31 Lyell/Secord, *Principles*, p. 196.
32 Ibid., p. 199.
33 Ibid., p. xxxiv, quoting Lyell to C. Lyell, 19 Oct 1830, Lyell (ed.), *Life, Letters and Journals*, Vol. 1, p. 308.
34 Lyell/Secord, *Principles*, p. xxx.

CHAPTER 5 FIGHTING FELLOWS

1 Lewis and Knell (eds), *Making of the Geological Society*, p. 77.
2 Ibid., p. 240.
3 Winchester, *Map That Changed the World*, p. 239.

4 Ibid., pp. 240–1.
5 Ibid., p. 37.
6 Cited in Herries Davies, *Whatever is Under the Earth*, p. 24.
7 Babbage cited in Lewis and Knell (eds), *Making of the Geological Society*, p. 327.
8 John C. Thackray, 'To See the Fellows Fight: Eye Witness Accounts of Meetings of the Geological Society of London and its Club', 1822–66, *British Society for the History of Science Monographs*, p. ix.
9 Martin J. S. Rudwick, *Worlds Before Adam*, Chicago: University of Chicago Press, 2008, pp. 31–2; see also Cadbury, *Dinosaur Hunters*, pp. 106–7.
10 O'Connor, *The Earth on Show*, p. 329.
11 Cadbury, *Dinosaur Hunters*, pp. 107–8.
12 Ibid., p. 108.
13 Ibid.
14 Lyell to Mantell, 8 Feb 1822, cited in O'Connor, *The Earth on Show*, p. 91, n. 66.
15 Conybeare to De la Beche, 4 March 1824, letter 302, T. Sharpe and P. J. McCartney, *The Papers of H. T. De la Beche (1796–1855) in the National Museum of Wales*, Cardiff: National Museums & Galleries of Wales, 1998, p. 33.
16 Lewis and Knell (eds), *Making of the Geological Society*, p. 323.
17 Lyell to Sisley, 21 Jan 1728, cited in ibid., p. 28.
18 Shelley Emling, *The Fossil Hunter*, New York: Palgrave Macmillan, 2009, p. 157.
19 Rachel Hewitt, *Map of a Nation: A Biography of the Ordnance Survey*, London: Granta, 2010, p. 29l.

CHAPTER 6 DATING THE DELUGE

1 Cadbury, *Dinosaur Hunters*, p. 21.
2 W. Buckland, '*Vindiciae geologicae*; or, the Connexion of Geology with Religion Explained', Oxford: 1820, pp. 2–5.
3 James Secord, *Victorian Sensation: The extraordinary publication, reception, and secret authorship of 'Vestiges of the Natural History of Creation'*, Chicago: University of Chicago Press, 2001, p. 223.
4 Buckland, '*Vindiciae geologicae*', pp. 2–5.
5 Ibid.
6 Ibid., pp. 23–4.

7 Ibid.
8 Mrs. E. O. Gordon, *The Life and Correspondence of William Buckland, D.D, F.R.S, by His Daughter*, London: John Murray, 1894, p. 4.
9 William Buckland, 'Memoir', in *Geology and Mineralogy Considered with Reference to Natural Theology*, London: William Pickering, 1837, Vol. 1, p. xxiv.
10 Cadbury, *Dinosaur Hunters*, p. 30.
11 O'Connor, *The Earth on Show*, p. 115.
12 Ibid., p. 80.
13 Ibid.
14 Holmes, *The Age of Wonder*, p. 447.
15 O'Connor, *The Earth on Show*, p. 182.
16 *The Times*, 23 Jun 1832, p. 4.
17 O'Connor, *The Earth on Show*, p. 75.
18 Cadbury, *Dinosaur Hunters*, p. 61.
19 William Wollaston to Buckland, 24 Jun 1822, quoted in O'Connor, *The Earth on Show*, p. 105.
20 Davy cited in O'Connor, *The Earth on Show*, p. 88; Buckland's letter to Lady Mary Cole, 24 Dec 1822, cited in Paul Ferris, *Gower in History: Myth, People, Landscape*, Hay on Wye: Armanaleg Books, 2009, p. 42.
21 Cited in McGowan, *The Dragon Seekers*, p. 58.
22 O'Connor, *The Earth on Show*, p. 91, n. 67, n. 68.
23 Ferris, *Gower*, p. 43.
24 Ibid., pp. 38–40.
25 Reverend William Buckland, 'Notice of the *Megalosaurus*, or Great Fossil Lizard of Stonesfield', *Transactions of the Geological Society of London* (1824), p. 391.
26 Lyell/Secord, *Principles*, p. 433.
27 Ibid., pp. 434–5.
28 Ibid., p. xxviii.
29 Ibid.
30 Richard Fortey, 'In Good Measure' / 'A Flood of Fossils', *TLS*, 2008 (source unknown).

CHAPTER 7 ON THE BEACH

1 John Fowles, *The French Lieutenant's Woman*, London: Vintage Books, 2010, p. 3.
2 Emling, *Fossil Hunter*, p. 55.

3 *Bristol Mirror*, 1823, cited in Donald R. Prothero, *The Story of Life in 25 Fossils: Tales of Intrepid Fossil Hunters*, New York: Columbian University Press, 2015, p. 168.
4 Emling, *Fossil Hunter*, p. 208.
5 Ibid., p. 143.
6 Ibid., pp. 115, 116, 182, 325–6.
7 Cited in Gordon, *Life and Correspondence of William Buckland*, p. 115.
8 Emling, *Fossil Hunter*, p. 193.
9 W. D. Lang, 'Three Letters of Mary Anning', *Proceedings of the Dorset Natural History and Archaeological Society*, Vol. 66 (1944), p. 171, cited in Lewis and Knell (eds), *Making of the Geological Society*, p. 339; Lang Papers, Dorset County Museum, cited in ibid., p. 101.
10 Lewis and Knell (eds), *Making of the Geological Society*, p. 197.
11 Ibid., p. 199.
12 Fowles, *French Lientenant's Woman*, p. 46.

CHAPTER 8 DINOSAUR WARS

1 Mantell's letters are preserved in the private archives of his son, Roderick, in Wellington, New Zealand.
2 Gideon Mantell, *Journal of Gideon Mantell, Surgeon and Geologist Covering the Years 1818–1852*, edited by E. Cecil Curwen, London: Oxford University Press, 1940, p. 3.
3 Dean, *James Hutton*, pp. 24–5.
4 Dennis R. Dean, *Gideon Mantell and the Discovery of Dinosaurs*, Cambridge: Cambridge University Press, 1999, p. 20.
5 Ibid., p. 38.
6 Ibid., p. 41.
7 Wilson, *Revolution in Geology*, p. 48.
8 McGowan, *The Dragon Seekers*, p. 87.
9 Emling, *Fossil Hunter*, p. 187.
10 Cited in Lewis and Knell (eds), *Making of the Geological Society*, p. 337.
11 McGowan, *The Dragon Seekers*, pp. 106–7.
12 Mantell, *Journal*, 17 Dec 1845, p. 200.
13 *The Times*, 3 Dec 1838; also see www.rth.org.uk.
14 Mantell, *Journal*, Dec 1845, p. 200.
15 Ibid., 4 and 23 Mar 1848, p. 221.

16 Ibid.
17 Ibid., p. 256.
18 Ibid., 25 Feb and 18 Mar 1851, pp. 264–5.
19 Ibid., 8 Oct 1852, pp. 273–4.
20 Ibid., 11 Oct 1851, p. 275.
21 Ibid., 13 Nov 1851, p. 277.
22 Bryson, *A Short History of Nearly Everything*, p. 118.
23 Mantell, *Journal*, 17 Sep 1829, p. 72.

CHAPTER 9 CELIBACY GALORE

1 Clark and Hughes (eds), *Life and Letters of the Reverend Adam Sedgwick*, Vol. 1, p. 183.
2 Ibid., p. 130.
3 Ibid., p. 152.
4 Ibid., p. 163.
5 Ibid., p. 161.
6 Ibid., p. 143.
7 Ibid., p. 433.
8 A prebendary is a clergyman receiving a stipend from a cathedral for having a role in its administration; a prebend is a form of benefit, usually from income on the cathedral estates.
9 Clark and Hughes (eds), *Life and Letters of the Reverend Adam Sedgwick*, Vol. 1, p. 452.
10 Ibid., p. 298.
11 Ibid., p. 299.
12 Ibid., p. 323.
13 Sedgwick cited in Adelene Buckland, *Novel Science: Fiction and the Invention of Nineteenth-Century Geology*, Chicago and London: University of Chicago Press, 2013, p. 87.
14 Clark and Hughes (eds), *Life and Letters of the Reverend Adam Sedgwick*, Vol. 1, p. 227.
15 Ibid., p. 247.
16 Ibid., p. 248.
17 Ibid., p. 211.
18 Source unknown.
19 Conybeare cited in Nicolaas Rupke, *The Great Chain of History: William Buckland and the English School of Geology (1814–1849)*, Oxford: Clarendon Press, 1983, p. 182.

20 Ibid.
21 Ibid.
22 Clark and Hughes (eds), *Life and Letters of the Reverend Adam Sedgwick*, Vol. 1, p. 275.
23 Rev. Adam Sedgwick, *Addresses delivered at the Anniversary Meeting of the Geological Society of London on the 18th February 1831 (and on the 19th February, 1830)*, London: Richard Taylor, 1831, p. 18.
24 Cited in Winchester, *Map That Changed the World*, pp. 280–1.
25 Clark and Hughes (eds), *Life and Letters of the Reverend Adam Sedgwick*, Vol. 1, p. 368.
26 Cited in ibid., p. 368.
27 Ibid.
28 Lyell/Secord, *Principles*, p. xxix.
29 Janet Browne, *Charles Darwin*, Vol. 2: *The Power of Place*, New York, Alfred Knopf, 2002, p. 140.
30 Clark and Hughes (eds), *Life and Letters of the Reverend Adam Sedgwick*, Vol. 1, p. 395.
31 Ibid.
32 Ibid., p. 459.
33 Ibid., p. 97.
34 Ibid., p. 460.
35 Lyell to his sister Caroline, 3 May 1837, Lyell (ed.), *Life, Letters and Journals*, Vol 2, p. 10.
36 Clark and Hughes (eds), *Life and Letters of the Reverend Adam Sedgwick*, Vol. 1, p. 388.
37 Ibid., Vol. 2, p. 60.
38 Ibid., Vol. 2, p. 136.
39 Lyell (ed.), *Life, Letters and Journals*, Vol. 1, p. 374.

CHAPTER 10 FROM SILURIA TO THE MOON

1 Archibald Geikie, *The Life of Sir Roderick K. Murchison*, London: John Murray, 1875, p. 96.
2 Rudwick, *Great Devonian Controversy*, pp. 73–8.
3 Ibid.
4 Robert A. Stafford, *Scientist of Empire: Sir Roderick Murchison, Scientific Exploration and Victorian Imperialism*, Cambridge: Cambridge University Press, 1989, p. 19.
5 Clark and Hughes (eds), *Life and Letters of the Reverend Adam Sedgwick*, Vol. 2, p. 442.

6 John C. W. Cope, 'What have they done to the Ordovician?', *Geoscientist*, 17 Mar 2007, Vol. 17, No 3.

CHAPTER 11 ALPS ON ALPS ARISE

1 Edward Whymper, *Scrambles Amongst the Alps in the Years 1860–69*, revised and edited by H. E. G. Tayndale, London: John Murray, 1879, p. 310.
2 Ibid., pp. 317–18.
3 Wordsworth, 'Cambridge and the Alps', *Complete Poetical Works*, Book VI, pp. 512–13.
4 *Poetical Register*, 1808, ii. 308, p. 311.
5 Whymper, *Scrambles Amongst the Alps*, p. 385.
6 Edward Whymper, letter to *The Times*, dated 7 Aug 1865, printed 8 Aug 1865.
7 Andrew St George to author, Jul 2011.
8 John Tyndall, *Fragments of Science: A Series of Detached Essays, Addresses, and Reviews*, London: Longmans, Green and Co., 1879, Vol. 2, p. 90.
9 Ibid., p. 128.
10 See John Stuart Mill, 'The Spirit of the Age', in *Essays on Politics and Culture*, edited by Gertrude Himmelfarb, New York: Doubleday, 1962, pp. 3–50.
11 Thomas Henry Huxley, *Collected Essays*, Vol. 1, first published 1894; digital publication: Cambridge University Press, 2011, p. 103.

CHAPTER 12 DARWIN THE GEOLOGIST

1 Charles Darwin to Catherine Darwin, 6 Apr 1834, *The Correspondence of Charles Darwin: 1821–1836*, Vol. 1, Cambridge: Cambridge University Press, 1985, p. 379.
2 Cited in Browne, *Darwin*, Vol. 2, p. 72.
3 Martin A. Patchett, *Life and letters of the Right Honourable Robert Lowe, Viscount Sherbrooke, G.C.B., D.C.L.*, London: Longmans, Green and Co., 1893, p. 20.
4 Ruth Padel, *Darwin: A Life in Poems*, London: Chatto & Windus, 2009, p. 31.
5 James Secord (ed.), *Charles Darwin: Evolutionary Writings (including The Autobiographies)*, Oxford: Oxford University Press, 2008, p. 81.
6 Darwin to Henslow, 12 Aug 1835, cited in John Bowlby, *Charles Darwin: A New Biography*, London: Hutchinson, 1990, p. 166.

7 Ibid.
8 Ibid., pp. 160–1.
9 Lyell to Darwin, 26 Dec 1836, in Lyell (ed.), *Life, Letters and Journals*, Vol. 1, p. 475.
10 Lyell to his sister Eleanor, 26 Feb 1830, in ibid., Vol. 1, p. 263.
11 Darwin to his brother, 6 Nov 1836, in ibid., Vol. 2.
12 Charles Lyell to Darwin, 26 December 1836, in ibid., Vol. 1, p. 475.
13 Lyell to Darwin, 26 Dec 1836, in ibid., Vol. 1, p. 475.
14 Desmond and Moore, *Darwin's Sacred Cause*, p. 257.
15 Thomas Pennant, *A Tour in Scotland and Voyage to the Hebrides*, Vol. 3, London: 1776, appendix, p. 394.
16 Lyell to Herschel, 1 Jun 1836, Lyell (ed.), *Life, Letters and Journals,* Vol. 1, p. 464.
17 Cited in Browne, *Darwin*, Vol. 2, p. 431.
18 Darwin to Lyell, 12 November 1838, *The Correspondence of Charles Darwin: 1837–1843*, Vol. 2, p. 114.
19 Emma Wedgwood to Darwin, in Browne, *Charles Darwin,* Vol. 1: *Voyaging*, London: Jonathan Cape, 1995, p. 396.
20 Wilson, *Lyell in America*, p. 287.
21 Darwin to Lyell, 6 September 1861, *The Correspondence of Charles Darwin: 1861*, Vol. 9, p. 256.
22 Darwin to Thomas Jamieson, 6 Sep 1861, in Ibid., p. 255.

CHAPTER 13 THE ICEMAN COMETH

1 Cited in Rupke, *Great Chain of History*, p. 102.
2 *Geoscientist*, Vol. 19, No. 2, p. 7.
3 Preface to Gordon, *The Life and Correspondence of William Buckland*, p. x.
4 Horace B. Woodward, *The History of the London Geological Society*, London: Longmans, Green, & Co., 1908, p. 140.
5 Forbes to Agassiz, 13 Feb 1841, quoted in Herries Davies, *Whatever is Under the Earth*, p. 87.
6 Edward Lurie, *Louis Agassiz: A Life in Science*, Chicago: University of Chicago Press, 1960, p. 101.
7 Cited in *Literary Gazette, and Journal of the Belles Lettres . . .*, London, Sat 17 Oct 1840, p. 671.
8 Herries Davies, *Whatever is Under the Earth*, p. 84.

9 Lyell/Secord, *Principles*, p. xi.
10 In *Louis Agassiz: His Life and Correspondence*, Vol. 2, Boston: Houghton Mifflin and Company, 1886, p. 445–6.
11 Henry Adams, *The Education of Henry Adams: An Autobiography*, Boston and New York: Houghton Mifflin Company, 1918, p. 60.
12 Oliver Wendell Holmes, 'At the Saturday Club', 1884.

CHAPTER 14 FOOTPRINTS IN PENNSYLVANIA

1 Darwin to the Reverend Fox, 24 Oct 1839, *The Correspondence of Charles Darwin: 1837–1843*, Vol. 2, p. 234.
2 Darwin to Leonard Jenyns, 10 Apr 1839, quoted in Francis Darwin (ed.), *Life and Letters of Charles Darwin*, London: John Murray, 1887, Vol. 1, p. 299.
3 Lyell to Darwin, Jul 1841, Wilson, *Revolution in Geology*, p. 460.
4 Darwin to Emma Wedgwood, 20 Jan 1838, in ibid., pp. 458–9.
5 C. Lyell, *Travels in North America; with Geological Observations on the United States, Canada and Nova Scotia*, London: John Murray, 1855, Vol. 1, p. 5.
6 Ibid., Vol. 1, p. 71.
7 Ibid., Vol. 1, pp. 60–1.
8 Ibid., Vol. 1, p. 17.
9 Ibid., Vol. 2, p. 106.
10 Dickens, *Bleak House*, London: Bradbury and Evans, 1853, Vol. 1, p. 1.
11 Lyell, *Travels*, Vol. 1, p. 102.
12 Ibid., Vol. 1, p. 121.
13 Ibid., Vol. 1, p. 105.
14 Ibid., Vol. 1, p. 206.
15 Ibid., Vol. 1, p. 107.
16 Ibid.
17 Ibid., Vol. 1, p. 117.
18 Ibid.
19 Ibid., Vol. 1, p. 158.
20 Ibid., Vol. 2, p. 66.
21 Ibid., Vol. 1, p. 168.
22 Ibid., Vol. 1, p. 169.
23 Ibid., Vol. 1, p. 183.
24 Ibid., Vol. 1, p. 228.
25 *Geoscientist*, Vol. 19, No. 2, p. 7.

26 Lyell, *Travels*, Vol. 1, p. 201.
27 Ibid., Vol. 1, p. 161.
28 Ibid., Vol. 1, p. 19.
29 Ibid., Vol. 2, pp. 62–3.
30 Lyell to his father, 25 Feb 1846, in Lyell (ed.), *Life, Letters and Journals*, Vol. 2, p. 101.
31 Lyell to Leonard Horner, Aug 1841, in ibid., p. 55.
32 Lyell to Leonard Horner, 15 Feb 1846, in ibid., p. 101.
33 Darwin to Lyell, 30 July–2 August 1845, *The Correspondence of Charles Darwin: 1844–1846*, Vol. 3, p. 233.
34 C. Lyell, *A Second Visit to the United States*, London: John Murray, 1850, Vol. 1, p. 184.
35 Ibid., p. 183.
36 Lyell, *Travels*, Vol. 2, p. 82.
37 Ibid., p. 13.
38 Ibid., p. 27.
39 Ibid., p. 15.
40 Ibid., p. 79.
41 Wilson, *Lyell in America*, p. 265.
42 Lyell, *Second Visit*, Vol. 2, p. 312.

CHAPTER 15 AT LAST, THE BIG QUESTION

1 Lyell/Secord, *Principles*, p. xxiv., citing Lyell to G. Ticknor, 1850, in Lyell (ed.), *Life, Letters and Journals*, Vol. 2, pp. 168–9.
2 Lyell (ed.), *Life, Letters and Journals*, Vol. 1, p. 313.
3 Wilson, *Lyell in America*, p. 301.
4 Ibid.
5 Desmond and Moore, *Darwin's Sacred Cause*, p. 354.
6 Lyell to Gideon Mantell, 24 Sep 1848, in Lyell (ed.), *Life, Letters and Journals*, Vol. 2, p. 148.
7 Desmond and Moore, *Darwin's Sacred Cause*, p. 394.
8 Source unknown.
9 Lyell, *Travels*, Vol. 1, p. 141.
10 Charles Lyell, *A Manual of Elementary Geology*, 4th edn, London: John Murray, 1852, p. xxii.
11 Lyell/Secord, *Principles*, p. 438.
12 Ibid.
13 Ibid., p. 233.

CHAPTER 16 ORIGIN OF *ORIGIN*

1 Cited in Winchester, *Map That Changed the World*, p. 113.
2 Cited in Cadbury, *Dinosaur Hunters*, p. 256.
3 Robert Chambers, *Vestiges of the Natural History of Creation*, London: J. Churchill, 1844, edited with an introduction by J. A. Secord, Chicago: University of Chicago Press, 1994, p. 277.
4 Sedgwick to M. Napier, 4 May 1845, quoted in Secord, *Victorian Sensation*, p. 240.
5 Proceedings of the Geological Society, Vol. 1, i., 1834, pp. 307–08; also Rupke, *Great Chain of History*, p. 178.
6 Cited in Secord, *Victorian Sensation*, p. 330.
7 *Science*, 26 Nov 2010, quoted in *The Times*, 26 Nov 2010, p. 35.
8 Cited in Lyell/Secord, *Principles*, introduction by Secord, pp. xxxviii-i1.
9 Clark and Hughes (eds), *Life and Letters of the Reverend Adam Sedgwick*, Vol. 2, p. 391.
10 Frank H. T. Rhodes, 'Darwin's search for a theory of the earth', *British Journal of the History of Science: Darwin and Geology*, Jun 1991, pp. 225ff.
11 Bowlby, *Darwin*, p. 307.
12 Ibid., p. 256.
13 Philip Henry Gosse, *Omphalos: An Attempt to Untie the Geological Knot*, London: John Van Voorst, 1857, p. 27.
14 Letter to Lyell, 3 May 1856, in *The Correspondence of Charles Darwin: 1856–1857*, Vol. 6, p. 100.
15 Browne, *Darwin*, Vol. 2, p. 17.
16 Ibid., p. 54.
17 Charles Darwin, *On the Origin of Species by Means of Natural Selection, or the Preservation of Favoured Races in the Struggle for Life*, London, John Murray, 1859, chapter 6, p. 228, 1900 edition. In chapter 3, p. 77, he credits Spencer for the expression 'survival of the fittest'.
18 Cited in Browne, *Darwin*, Vol. 2, p. 80.
19 Ibid., p. 100.
20 Ibid.
21 Ibid., p. 669.
22 Ibid., pp. 669–70.
23 Ibid., p. 94.
24 Ibid.
25 Ibid.

26 Clark and Hughes (eds), *Life and Letters of the Reverend Adam Sedgwick*, Vol. 2, p. 360.
27 Quoted in Bowlby, *Darwin*, p. 367.
28 Darwin to Hooker, 12 September 1847, *The Correspondence of Charles Darwin: 1847–1850*, Vol. 4, p. 74.
29 Tyndall, *Fragments of Science*, Vol. 1, London 1879, p. 207.
30 Ibid., p. 211.
31 Ibid.
32 Darwin to Thomas Jamieson, 6 September 1861, *The Correspondence of Charles Darwin: 1861*, Vol. 9, p. 255.

CHAPTER 17 THE WHOLE ORANG

1 Lyell to Darwin, 4 November 1864, Lyell (ed.), *Life, Letters and Journals*, Vol. 2, p. 384.
2 Lyell to Darwin, 16 Jan 1865, Lyell (ed.), *Life, Letters and Journals*, Vol. 2, p. 384.
3 Ibid., p. 330.
4 Lyell to George Ticknor, 9 Jan 1860, in ibid., p. 328–9.
5 Ibid.
6 Lyell to Horner, 26 Dec 1861, in ibid., p. 353.
7 Browne, *Darwin*, Vol. 2, p. 218.
8 Lyell (ed.), *Life, Letters and Journals*, Vol. 2, p. 363.
9 Ibid., p. 219, in Browne, *Darwin*, Vol. 2
10 Ibid.
11 Lyell to Darwin, 5 May 1869, in Lyell (ed.), *Life, Letters and Journals*, Vol. 2, p. 442.
12 Charles Lyell, *Geological Evidences of the Antiquity of Man*, London: John Murray, 1863, p. 506.
13 Darwin to Hooker, 25 Feb 1863, *The Correspondence of Charles Darwin: 1863*, Vol. 11, pp. 173–4.
14 Lyell to Hooker, 9 Mar 1863, in Lyell (ed.), *Life, Letters and Journals*, Vol. 2, p. 316.
15 Desmond and Moore, *Darwin's Sacred Cause*; p. 495.
16 Ibid., p. 496.
17 Frank James, 'Science and Religion', *London Library Magazine*, summer 2011, p. 16.
18 Desmond and Moore, *Darwin's Sacred Cause*, p. 495.
19 Browne, *Darwin*, Vol. 2, p. 115.
20 Ibid., Vol. 1, p. 387.

21 Lyell (ed.), *Life, Letters and Journals*, Vol. 2, p. 406 and Appendix C, p. 471.
22 Hooker's eloquent words appeared in the *Manchester Guardian* on 3 March 1875.
23 Browne, *Darwin*, Vol. 2, p. 248.
24 Ibid.
25 Ibid., p. 417.
26 Desmond and Moore, *Darwin's Sacred Cause*, p. 614; Browne, *Darwin*, Vol. 2, p. 481.
27 *Nature*, 4 Mar 1875; Lyell (ed.), *Life, Letters and Journals*, Vol. 2, p. 475.
28 Lyell to Spedding, 19 May 1863, Lyell (ed.), *Life, Letters and Journals*, Vol. 2, p. 374.

CHAPTER 18 MUSEUM PIECES

1 Cited in Browne, *Darwin*, Vol. 2, p. 418.
2 *The Times*, 18 Apr 1881.
3 Emling, *Fossil Hunter*, p. 208.
4 Clark and Hughes (eds), *Life and Letters of the Reverend Adam Sedgwick*, Vol. 2, p. 475.
5 Ibid., p. 177.
6 Preface by the Reverend Adam Sedgwick, to J. W. Salter, Adam Sedgwick, John Morris, *A Catalogue of the Collection of Cambrian and Silurian Fossils*, New York: Cambridge University Press, 1873, p. xxxi.
7 Winchester, *Map That Changed the World*, p. 218.

CHAPTER 19 THEN AND NOW

1 Richard Fortey, *Trilobite! Eyewitness to Evolution*, London: Random House, 2001, p. 117.
2 'Misadventures in the Burgess Shale', Desmond Collins, *Nature*, Vol. 460, 20 Aug 2009, pp. 952–3. (http://www.nature.com/nature/journal/v460/n7258/full/460952a.html?message=remove&FORM=ZZNR4).
3 *Nature*, 1 Jul 2010, pp. 41–2.
4 Ibid., 17 Feb 1870.
5 See https://www.geolsoc.org.uk/Geoscientist/Archive/June-2011/Holmess-first-date.
6 *Nature*, 28 Apr 2005, Vol. 434, p. 1053.
7 See Norman Cohn, *Noah's Flood: The Genesis Story in Western Thought*, New Haven: Yale, 1996, p. 128.

8 *Nature*, vol. 467, 9 September 2010, p. 150 (http://www.nature.com/news/2010/100908/full/467150a.html?s=news_rss)
9 Stephen Hawking, *The Times*, 'Eureka', 3 Sep 2010, p. 3.
10 Arthur Holmes, *Principles of Physical Geology*, New York: The Ronald Press Company, 1945, p. 496.
11 Ibid.
12 David Oldroyd, 'Held in Place by Practice', *Science*, Vol. 284, 16 Apr 1999, p. 440.
13 Ted Nield, *Supercontinent: Ten Billion Years in the Life of Our Planet*, Cambridge, MA: Harvard University Press, 2009, p. 15.
14 *National Geographic*, 4 Mar 2010.
15 Review of Nield's *Supercontinent*, *Guardian*, 6 Oct 2007.
16 Fortey, p. 34.

SELECT BIBLIOGRAPHY

Baker, T. C., and C. I. Savage, *An Economic History of Transport in Britain*, London: Hutchinson, 1959, revised 1974

Baker, Victor R., 'Catastrophism and uniformitarianism: logical roots and current relevance in geology', London: Geological Society, Special Publications 1998, v. 143, pp. 171–82

Bakewell, Robert, *An Introduction to Geology, Illustrative of the General Structure of the Earth, Comprising the Elements of the Science, and an Outline of the Geology and Mineral Geography of England*, London: J. Harding, 1813

Ball, Philip, *Life's Matrix: A Biography of Water*, New York: Farrar, Straus and Giroux, 2000

Balzac, Honoré de, *The Wild Ass's Skin* (*La Peau de chagrin*, 1831), London: Everyman Paperback, 1967

Barber, Lynn, *The Heyday of Natural History*, London: Cape, 1980

Barrett, Paul H., 'The Sedgwick-Darwin Geologic Tour of North Wales', *Proceedings of the American Philosophical Society*, Vol. 118, No. 2, April 1974, pp. 146–64

Baxter, Stephen, *Ages in Chaos: James Hutton and the Discovery of Deep Time*, London: Weidenfeld & Nicolson, 2003

Beer, Gillian, *Darwin's Plots: Evolutionary Narrative in Darwin, George Eliot and Nineteenth-Century Fiction*, London: Routledge & Kegan Paul, 1983

Bentley, Christopher, *A Summer of Hummingbirds*, New York: Penguin, 2008

Bompas, George C. Thomas, *Life of Frank Buckland by his Brother-in-Law*, London: Smith, Elder & Co., 1885

Bowlby, John, *Charles Darwin: A New Biography*, London: Hutchinson, 1990

Brock, M. G. and M. C. Curthoys, 'Oxford's Scientific Awakening and the Role of Geology', in M. G. Brock and M. C. Curthoys (eds), *The History of the University of Oxford*, Part I, Oxford: Clarendon Press, 2000, pp. 543–62

Browne, Janet, *Charles Darwin*, Vol. 1: *Voyaging*, London: Jonathan Cape, 1995

—*Charles Darwin,* Vol. 2: *The Power of Place*, New York: Alfred Knopf, 2002

Bryson, Bill, *A Short History of Nearly Everything*, London: Black Swan, 2003

Buckland, the Reverend William, B.D., F.R.S. F.L.S., '*Vindiciae geologicae*; or the Connexion of Geology with Religion Explained', an inaugural lecture delivered before the University of Oxford, 15 May 1819, on the endowment of a Readership in Geology by His Royal Highness, the Prince Regent, Oxford: 'at the University Press for the Author', 1820

—*Reliquiae Diluvinae; or, Observations on the Organic Remains contained in Caves, Fissures, and Diluvial Gravel, and on Other Geological Phenomena, attesting the Action of an Universal Deluge*, London: John Murray, 1st edn 1823, 2nd edn 1824

—'Notice of the *Megalosaurus*, or Great Fossil Lizard of Stonesfield', *Transactions of the Geological Society of London*, 1824, pp. 390–6

—'Antediluvian human remains', *American Journal of Science*, 18, 1830, pp. 393–4

—*Geology and Mineralogy Considered with Reference to Natural Theology*, London: William Pickering, 1st edn, 2 vols, 1836; 2nd edn, 2 vols, 1837

—*Bridgewater Treatise: Geology and Mineralogy*, 2 vols, London: William Pickering, 1836

Buffon, Georges, *Les Epoques de la Nature*, Paris: 1778

Burgess, G. H. O., *The Curious World of Frank Buckland*, London: John Baker, 1967

Cadbury, Deborah, *The Dinosaur Hunters*, London: Fourth Estate, 2000

Chambers, Robert, *Vestiges of the Natural History of Creation,* London: J. Churchill, 1844, edited with an introduction by J. A. Secord, Chicago: University of Chicago Press, 1994

Chevalier, Tracy, *Remarkable Creatures*, London: HarperCollins, 2009

Clark, J. W. and T. M. Hughes (eds), *The Life and Letters of the Reverend Adam Sedgwick*, 2 vols, London: C. J. Clay and Sons for Cambridge University Press, 1890

Cohn, Norman, *Noah's Flood: The Genesis Story in Western Thought*, New Haven: Yale, 1996

Collins, Desmond, 'Misadventures in the Burgess Shale', *Nature*, Vol. 460/20, August 2009

Conybeare, William, 'Notice of the discovery of a new Fossil Animal, forming a link between the *Ichthyosaurus* and the Crocodile, together with general remarks on the Osteology of the *Icthyosaurus*', *Transactions of the Geological Society of London*, 1821

—and William Phillips, *Outlines of the geology of England and Wales, with an introductory compendium of the general principles of that science, and comparative views of the structure of foreign countries*, Part I, London: William Phillips, 1822

Cope, John C. W., 'What have they done to the Ordovician?', *Geoscientist*, 17 March 2007, Vol. 17, No. 3, p. 19

Curwen, E. C., *Notes and introduction to The Unpublished Journal of Gideon Mantell: Surgeon and Geologist 1819–1852*, 8 vols, London: Oxford University Press, H. Milford, 1940

Cuvier, Georges, *Recherches sur les ossemens fossiles de quadrupèdes, où l'on rétablit les caractères de plusieurs espèces d'animaux que les révolutions du globe paroissent avoir détruites*, 4 vols, Paris: Deterville, 1812

—and Alexandre Brongniart, *Essai sur la géographie minéralogique des environs des Paris*, Paris: Baudouin, 1811

Darwin, Charles, *The Structure and Distribution of Coral Reefs*, London: Smith, Elder, 1842

—*Journal of Researches into the Natural History and Geology of the Countries Visited During the voyage of HMS Beagle*, London: J. Murray, 2nd edn, 1845, reprinted as *The Voyage of the Beagle*, ed. H. G. Cannon, London: J. M. Dent, 1959

—*The Complete Work of Charles Darwin Online*, including Darwin's Beagle field notebooks (1831–36), http://darwin-online.org.uk

—*On the Origin of Species by Means of Natural Selection, or the Preservation of Favoured Races in the Struggle for Life*, London: John Murray, 1859

Davies, Dr Margaret, *Victorian Naturalists in Tenby*, Tenby: Tenby Museum and Art Gallery, 1998

Davy, Humphry, *Consolations in Travel*, London: John Murray, 1830

Dawkins, Richard, *Science and Faith, The Seventh Athenaeum Lecture*, 30 September 2004

—*The God Delusion*, London: Bantam Books, 2006

Dean, Dennis R., *James Hutton and the History of Geology*, Ithaca, New York: Cornell University Press, 1992

—*Gideon Mantell and the Discovery of Dinosaurs*, Cambridge: Cambridge University Press, 1999

Desmond, Adrian and James Moore, *Darwin's Sacred Cause: Race, Slavery and the Quest for Human Origins*, London: Allen Lane, 2009

Dickens, Charles, *Bleak House*, London: Penguin Classics, 2003

Emling, Shelley, *The Fossil Hunter*, New York: Palgrave Macmillan, 2009

Endersby, Jim, 'Kew gooseberries', review of Vols 11, 12 and 13 of *The Correspondence of Charles Darwin, Times Literary Supplement*, 21 November 2003, pp. 3–4

—'Creative Designs', *Times Literary Supplement*, 16 March 2007

Ferris, Paul, *Gower in History: Myth, People, Landscape*, Hay on Wye: Armanaleg Books, 2009

Fortey, Richard, *Life: A Natural History of the First Four Billion Years of Life on Earth*, London: Harper Flamingo, 1998

—'Most Curious of Seas', *London Review of Books* (review of *Noah's Flood: The New Scientific Discoveries about the Event that Changed History*), 1 July 1999

—*Trilobite!: Eyewitness to Evolution*, London: Random House, 2001

—*The Earth: An Intimate History*, London: HarperCollins, 2004

—'In Retrospect: Leibniz's *Protogaea*', *Nature*, Vol. 455, 4 September 2008, p. 35

—*Dry Store Room No. 1: The Secret Life of the Natural History Museum*, London and New York: Harper Perennial, 2008

—*The Hidden Landscape: Journey into the Geological Past*, London: Bodley Head, 2010

Geikie, Archibald, *The Life of Sir Roderick K. Murchison, Based on his Journals and Letters with Notices of his Scientific Contemporaries and a Sketch of the Rise and Growth of Palaeozoic Geology in Britain*, 2 vols, London: John Murray, 1875

Gillispie, Charles Coulton, *Genesis and Geology: The Impact of Scientific Discoveries upon Religious Beliefs in the Decades before Darwin*, New York: Harper & Row 1951; HarperTorchbook, 1959

Gold, Thomas, 'The deep, hot biosphere', *Proceedings of the National Academy of Sciences*, USA, Vol. 89, pp. 6045–9, July 1992

Gordon, Mrs. E. O., *The Life and Correspondence of William Buckland, D.D, F.R.S, by his Daughter*, London: John Murray, 1894

Gorst, Martin, *Aeons: The Search for the Beginning of Time*, London: Fourth Estate, 2001

Gosse, Philip Henry, *A Naturalist's Rambles on the Devonshire Coast*, London: John Van Voorst, 1853

—*Tenby: A Sea-side Holiday*, London: John Van Voorst, 1856
—*Omphalos: An Attempt to Untie the Geological Knot*, London: John Van Voorst, 1857
Gould, Stephen Jay, *Time's Arrow, Time's Cycle: Myth and Metaphor in the Discovery of Geological Time*, Cambridge, Mass.: Harvard University Press, 1987
—*Dinosaur in a Haystack: Reflections on Natural History*, London: Cape, 1996
Greene, Mort, *Geology in the Nineteenth Century*, Ithaca, NY: Cornell University Press, 1982
Hawking, Stephen, *A Brief History of Time*, New York: Bantam Dell, 1998
—with Leonard Mlodinow, *The Grand Design*, New York, Bantam, 2010
Hawley, Duncan, '"The first true Silurian": an evaluation of the site of Murchison's discovery of the Silurian', *Proceedings of the Geologists' Association*, 108, pp. 131–40
Herbert, Sandra, 'Charles Darwin as a prospective geological author', *British Journal for the History of Science*, June 1991, pp. 159–92
—*Charles Darwin, Geologist*, Ithaca, New York: Cornell University Press, 2005
Herries Davies, Gordon L., *Whatever is Under the Earth: The Geological Society of London 1807–2007*, London: Geological Society, 2007
Hewitt, Rachel, *Map of a Nation: A Biography of the Ordnance Survey*, London: Granta, 2010
Holmes, Arthur, *The Age of the Earth*, London and New York: Harper & Brothers, 1913
—*Principles of Physical Geology*, New York: The Ronald Press Company, 1945
Holmes, Richard, *Coleridge: Early Visions*, London: Hodder and Stoughton, 1989
—*The Age of Wonder: How the Romantic Generation Discovered the Beauty and Terror of Science*, London: HarperPress, 2008
Howells, Sid, 'Geology & scenery of the Little Haven area', July 2009, unpublished
Hughes, C. P., 'The Ordovician trilobite faunas of the Builth-Llandrindod Inlier, Central Wales, *Bulletin of the British Museum (Natural History) Geology*, Part I: Vol. 18, No. 3, London, 1969; Part II: Vol. 20, No. 4, London, 1971; Part III, Vol. 32, No. 3, London, 27 September 1979
Hutton, James, *Theory of the Earth*, Hafner Press, 1795

— *Theory of the Earth*, 1788, published in *Transactions of the Royal Society of Edinburgh*, I, pp. 209–304; also *Theory of the Earth with Proofs and Illustrations*, Edinburgh: William Creech, 1795

Inkster, Ian, and Jack Morrell, *Metropolis and Province: Science in British Culture*, 1780–1850, London: Hutchinson, 1983

Jackson, Lee, *A Dictionary ofVictorian London: An A-Z of the Great Metropolis*, Anthem Art and Culture, London: Anthem Press, 2006

James, Frank A. J. L., *Michael Faraday: A Very Short Introduction*, Oxford: Oxford University Press, 2010

—and Margaret Ray, 'Science in the Pits: Michael Faraday, Charles Lyell and the Home Office Enquiry into the Explosion at Haswell Colliery, County Durham in 1844', *History and Technology*, 1999, Vol. 15, pp. 325–51

John, Dr Brian S., *The Geology of Pembrokeshire*, Aberteifi, Ceredigion: Abercastle Publications, 1979, republished 2003

Judd, John W., *The Coming of Evolution: The Story of a Great Revolution in Science*, Cambridge: Cambridge University Press, 1910

—*The Complete Work of Charles Darwin Online*, Cambridge University Press (http://darwin-online.org.uk)

Lamarck, Jean Baptiste, *Philosophie Zoologique, ou Exposition des considérations relatives à l'histoire naturelle des animaux*, Paris: 1809

Lang, W. D., 'Three letters by Mary Anning', *Proceedings of the Dorset Natural History and Archaeological Society*, Vol. 66, 1944, cited in C. L. E. Lewis and S. J. Knell (eds), *The Making of the Geological Society of London*, London: Geological Society, 2009, p. 339

Lewes, George Henry, *Sea-Side Studies*, Edinburgh: William Blackwood & Sons, 1858

Lewés, C. L. E. and S. J. Knell (eds), *The Making of the Geological Society of London*, London: Geological Society, 2009

Lightman, Bernard, *Victorian Popularizers of Science: Designing Nature for New Audiences*, Chicago: University of Chicago Press, 2007

Lurie, Edward, *Louis Agassiz: A Life in Science*, Chicago: University of Chicago Press, 1960

Lyell, Charles, 'Memoir on the Geology of Central France; including the Volcanic Formations of Auvergne, the Velay, and the Vivarais, with a Volume of Maps and Plates', by G. P. Scrope, FRS, FGF, London: *Quarterly Review*, Vol. 36, 1827

—*Elements of Geology*, London: John Murray, 1838; 2nd edn, 2 vols, 1841; 1 vol., 5th edn, 1861

—*A Manual of Elementary Geology*, 1 vol., lst edn, London: John Murray, 1838; 2 vols, 2nd edn, July 1841; 1 vol., 3rd edn, 1851; 4th edn, 1852
—*Principles of Geology*, and *Principles of Geology or the Modern Changes of the Earth Considered as Illustrative of Geology*, 2 Vols, London: John Murray, 1840
—'Royal Institution', *Literary Gazette and Journal of the Belles Lettres*, 12 February 1848, pp. 119–21
—'Royal Institution', *The Athenaeum*, 12 February 1848, pp. 66–7
—*A Second Visit to the United States*, 2 vols, London: John Murray, 1850
—*Travels in North America: with Geological Observations on the United States, Canada and Nova Scotia*, 2 vols, London: John Murray, 1855
—*Geological Evidences of the Antiquity of Man*, 2 vols, London: John Murray, 1863 and 1873
—*Principles of Geology*, edited with an introduction by James A. Secord, London: Penguin Classics, 1997
Lyell, Katherine Murray (ed.), *Life, Letters and Journals of Sir Charles Lyell*, 2 vols, London: John Murray, 1881
McCartney, P. J., *Henry De La Beche: Observations on an Observer*, Cardiff: Friends of the National Museum of Wales, 1977
McGowan, Christopher, *The Dragon Seekers: How an Extraordinary Circle of Fossilists Discovered the Dinosaurs and Paved the Way for Darwin*, Cambridge, Mass.: Perseus, 2001
MacLeod, Roy M., 'Scientific Advice for British India: Imperial Perceptions and Administrative Goals 1898–1923', *Modern Asian Studies*, 1975, pp. 343–84
Maddox, Bruno, 'Darwin's Great Blunder – and Why It Was Good for the World', *Discover*, November 2009
Maddox, John, 'The Age of the Earth', *Guardian*, 28 March 1961, p. 6
Mantell, Gideon, *The Fossils of the South Downs; or Illustrations of the Geology of Sussex*, London: Lupton Belfe, 1822
—*The Journal of Gideon Mantell, Surgeon and Geologist Covering the Years 1818–1852*, edited by E. Cecil Curwen, London: Oxford University Press, 1940
Miller, Hugh, *The Old Red Sandstone* (1841), London and Edinburgh: 20th edition, 1875; Boston: Gould and Lincoln, 1851
Mompas, G. C., *Life of Frank Buckland by his Brother-in-Law*, 6th edn, London: Smith, Elder, & Co., 1885
Morrell, J. B., and Arnold Thackray, *Gentlemen of Science. Early Years of the British Association for the Advancement of Science*, Oxford, Clarendon, 1981

Morris, Richard, *Time's Arrows: Scientific Attitudes Toward Time*, New York: Simon and Schuster, 1985

Murchison, Roderick I., *Silurian System: Founded on Geological Researches in the Counties of Salop, Hereford, Radnor, Montgomery, Caermarthen, Brecon, Pembroke, Monmouth, Gloucester, Worcester, and Stafford; with Descriptions of the Coalfields and Overlying Formations*, 2 vols, London: John Murray, 1839

—*Siluria*: *The History of the Oldest Known Rocks Containing Organic Remains, with a Brief Sketch of the Distribution of Gold Over the Earth*, London: John Murray, 1854

Nield, Ted, *Supercontinent: Ten Billion Years in the Life of Our Planet*, Cambridge, MA: Harvard University Press, 2009

O'Connor, Ralph, *The Earth on Show: Fossils and the Poetics of Popular Science, 1802–1856*, London and Chicago: University of Chicago Press, 2007

Oldroyd, David, 'Held in Place by Practice', review of Naomi Orestes, *The Rejection of Continental Drift Theory and Method in American Earth Science*, New York: Oxford University Press, 1999, in *Science*, 16 April 1999

Open University, *Geological Society Journal*, Symposium Edition, 2008, Volume 29 (2)

Owens, R. M., *Trilobites in Wales*, Cardiff: National Museum of Wales, 1984

Playfair, John, *Illustrations of the Huttonian Theory of the Earth*, Edinburgh: Cadell and Davies, 1802

—*The Works of John Playfair, with a memoir of the Author*, 4 volumes, Edinburgh: Archibald Constable and Co., 1822

Plimer, Ian, *Telling Lies for God: Reason vs. Creationism*, Sydney, Australia: Random House, 1994

Porter, Roy, *The Making of Geology: Earth Science in Britain, 1660–1815*, Cambridge and New York: Cambridge University Press, 1977

—'Gentlemen and Geology: the Emergence of a Scientific Career, 1660–1920', *The Historical Journal*, 21.4, December 1978, pp. 809–36

Redfern, Ron, *Origins: The Evolution of Continents, Oceans and Life*, London: Weidenfeld and Nicolson, 2000

Repcheck, Jack, *The Man Who Found Time: James Hutton and the Discovery of the Earth's Antiquity*, Cambridge, Mass.: Perseus, 2003

Rhodes, Frank H. T., 'Darwin's search for a theory of the earth', *British Journal of the History of Science: Darwin and Geology*, June 1991

Richet, Pascal, *A Natural History of Time*, translated by John Venerella, Chicago: University of Chicago Press, 2007

Rudwick, Martin J. S., 'Charles Lyell Speaks in the Lecture Theatre', *British Journal for the History of Science*, Lyell Centenary Issue, July 1976, Vol. 9, No. 37

—*The Great Devonian Controversy*, Chicago: University of Chicago Press, 1985

—*Scenes from Deep Time: Early Pictorial Images of the Prehistoric World*, Chicago: University of Chicago Press, 1992

—*Georges Cuvier, Fossil Bones, and Geological Catastrophes*, Chicago: University of Chicago Press, 1997

—*Bursting the Limits of Time*, Chicago: University of Chicago Press, 2005

—*Lyell and Darwin, Geologists: Studies in the Earth Sciences in the Age of Reform*, Aldershot: Ashgate Variorum, March 2005

—*Worlds Before Adam*, Chicago: University of Chicago Press, 2008

Rupke, Nicolaas A., *The Great Chain of History: William Buckland and the English School of Geology (1814–1849)*, Oxford: Clarendon Press, 1983

—*Richard Owen: Victorian Naturalist*, New Haven: Yale, 1994

Ryan, William and Walter Pitman, *Noah's Flood: The New Scientific Discoveries about the Event that Changed History*, New York: Simon and Schuster, 1999

Schofield, Robert E., *The Lunar Society of Birmingham: A Social History of Provincial Science and Industry in Eighteenth-Century England*, Oxford: Clarendon Press, 1963

Scrope, George Poulett, *Memoir on the Geology of Central France, Including the Volcanic Formations of Auvergne, the Velay and the Vivais*, 2 vols, London: Longman, Rees, Orme, Brown, and Green, 1827

Secord, Anne, 'Botany on a Plate: Pleasure and the Power of Pictures in Promoting Early Nineteenth-Century Scientific Knowledge', *Isis*, 93, pp. 28–37

Secord, James A., 'King of Siluria: Roderick Murchison and the Imperial Theme in Nineteenth-Century Geology', *Victorian Studies*, 25 (4), Summer, 1982, pp. 413–42

—*Controversy in Victorian Geology: The Cambrian-Silurian Dispute*, Princeton, NJ: Princeton University Press, 1986

—'The Geological Survey of Great Britain as a research school, 1839–1855', *History of Science*, 24, 1986, pp. 223–75

—'The discovery of a vocation: Darwin's early geology', *The British Journal of the History of Science*, June 1991, pp. 133–57

—*Victorian Sensation: The extraordinary publication, reception, and secret authorship of 'Vestiges of the Natural History of Creation'*, Chicago: University of Chicago Press, 2001

—(ed.), *Charles Darwin: Evolutionary Writings (including The Autobiographies)*, Oxford: Oxford University Press, 2008

Sedgwick, A., 'Address to the Geological Society', *Proceedings of the Geological Society of London*, Vol. 1, London: 1831, pp. 281–316

Sharpe, T., and P. J. McCartney, *The Papers of H. T. De la Beche (1796–1855) in the National Museum of Wales*, Cardiff: National Museums & Galleries of Wales, 1998

Sillitoe, Paul, 'The Role of Section H at the British Association for the Advancement of Science in the History of Anthropology', *Durham Anthropological Journal*, Vol. 13, 2

Sommer, Marianne, *Bones and Ochre: The Curious Afterlife of the Red Lady of Paviland*, Cambridge, Mass.: Harvard University Press, 2007

Stafford, Robert A., *Scientist of Empire: Sir Roderick Murchison, Scientific Exploration and Victorian Imperialism*, Cambridge: Cambridge University Press, 1989

Sutton, Mark D. et al., 'Silurian brachiopods with soft-tissue preservation', letter to *Nature*, 18 August 2005, 436, pp. 1013–15

Tattersall, Ian, *The Fossil Trail: How We Know What We Think We Know About Human Evolution*, Oxford: Oxford University Press, 1995

Thackray, John C., 'To See the Fellows Fight: Eye Witness Accounts of Meetings of the Geological Society of London and its Club', 1822–66, *British Society for the History of Science Monographs*, 12, 2003

Thwaite, Ann, *Glimpses of the Wonderful: The Life of Philip Henry Gosse*, London: Faber and Faber, 2002

Uglow, Jenny, *The Lunar Men of Birmingham*, London: Faber and Faber, 2002

Whewell, William, *Quarterly Review*, review of volume 2 of Lyell's *Principles of Geology*, 1830–33, 47, 1832, pp. 103–32

Wilson, Leonard G., *Charles Lyell: The Years to 1841: The Revolution in Geology*, New Haven and London: Yale University Press, 1972

—*Lyell in America: Transatlantic Geology, 1841–53*, New Haven and London: Yale University Press, 1998

Winchester, Simon, *The Map That Changed the World*, London: Viking, 2001; Penguin, 2002

Woodmorappe, John, *Studies in Flood Geology*, Institute for Creation Research, www.icr.org

Wordsworth, William, *The Excursion*, London: Edward Moxon, 1836

Worsley, Peter, 'Rocks of ages', *Geoscientist*, Vol. 18, No. 11, pp. 20–3

Wyse Jackson, Patrick, *The Chronologers' Quest: The Search for the Age of the Earth*, Cambridge: Cambridge University Press, 2006

ACKNOWLEDGEMENTS

Perhaps the best part of researching Victorian geologists was the Celtic field trips that took me to Glen Roy in the Scottish Highlands, the Pembrokeshire coast in Wales, the Llanfair Quarry near Builth Wells and Tricket Mill and the Llanstephan Hills on the banks of the River Wye. The subject also gave me the chance to work in the beautiful library of the Geological Society in Burlington House, Piccadilly, the society's home since 1860, with its remarkable collection of journals going back to the society's origins in 1807. At Burlington House I was grateful for the help of the assistant librarian, Wendy Cawthorne, who knows so well the vast collection.

Well before I went on the Glen Roy field trip led by Professor Martin Rudwick FRS, I had been inspired by his fine books on the history of the earth sciences – notably *Worlds Before Adam*, *Bursting the Limits of Time*, *Lyell and Darwin, Geologists: Studies in the Earth Sciences in the Age of Reform* and *The Great Devonian Controversy.* On the Glen, I met Professor James A. Secord, whose Penguin classic edition of Charles Lyell's *Principles of Geology* never left my desk while writing this book. I also met John and Annie Henry of Nineteenth-Century Geological Maps, who offered many illustrations from Lyell's archives for this book as well as helpful comment on the text.

At Little Haven, the Pembrokeshire geologist Sid Howells showed me round the spectacular folds, twists and caves, bringing me a hardhat to protect me from falling rocks. None fell.

Dr Robert Owens, head of the palaeontology section of the National Museum of Wales at Cardiff, took me around the fossiliferous Llanfawr Quarry at Llandrindod Wells. Trilobites may be the world's favourite fossil – hard carapaced creatures with fringed heads. As one who had never seen a trilobite before, I came home with pockets full – not museum-worthy specimens, rather fragments found in almost every shard of scattered rock. From Tom Sharpe, curator of palaeontology and archives and Dave Smith at the same fine museum, I learned of the great collection of the papers and drawings of Sir Henry Thomas De la Beche. I am grateful to the museum for permission to reproduce some of these classics.

Charles Gordon Clark of Llaneglwys, Powys, escorted me on a memorable field trip to the iconic juncture of the Old Red Sandstone with the Aberedw rocks at Llanstephan in the Upper Wye Valley. He also gave much help, with accompanying literature, on the long dispute to define the Silurian System of rocks. Professor Jim Kennedy and Eliza Howell, Geological Collections Manager, showed me round the Buckland collection at the Ashmolean Museum, Oxford.

The Natural History Museum in South Kensington is the professional home of the man with possibly the greatest gift for explaining geology, Professor Richard Fortey, FRS, FRSL. The title of his book *Trilobite!* conveys an infectious enthusiasm for the tiny three-lobed fossils which existed before life emerged from the sea. His subtitle calls them *Eyewitness to Evolution*.

Once more I am grateful for the assistance of Dr Walter Gratzer, emeritus professor of biophysical chemistry at King's College London and family friend. His rare mix of scientific and literary fluency is shown in his many books, such as *Eurekas and Euphorias: The Oxford Book of Scientific Anecdotes* and, most recently, *Giant Molecules: From Nylon to Nanotubes*. He kindly read the manuscript of this book and made many suggestions for changes, small and large, almost all of them accepted. Another friend, the geologist Dr Laura Garwin, cast her expert eye on the text and offered many helpful observations. Again I owe thanks to the scholar Bernard McGinley for casting his sharp eye over another of my works in progress. Any errors that remain are my own.

I would also like to thank Aosaf Asfal of the Royal Society, Sally Bushell of the University of Lancaster, Dr Warwick Gould of the University of London, Frank A. J. L. James, professor of the history of science at the Royal Institution, the geological historian Cherry Lewis, the science writer Nina Morgan, Gwyn Miles of Somerset House, Paul Shepherd of the British Geological Survey at Keyworth, Nottingham, James and Annie Secord and Hugh Torrens of Keele University – all of whom were generous with their time and explanations for an outsider. Sian Williams, my assistant, gave invaluable technical support.

As ever, I am aware of my good fortune in having, as literary agents and friends, Caradoc King of A. P. Watt Ltd. in London and Ellen Levine of the Ellen Levine Agency in New York.

And as always I am deeply grateful to my family – my daughter Bronwen, my granddaughter Laura and my son Bruno, all united in loving memory of my husband, Sir John Maddox, former editor of *Nature*, who died in April 2009 after I began this book and who encouraged me to write it.

INDEX

A NOTE ON THE TYPE

The text of this book is set in Linotype Sabon, a typeface named after the type founder, Jacques Sabon. It was designed by Jan Tschichold and jointly developed by Linotype, Monotype and Stempel in response to a need for a typeface to be available in identical form for mechanical hot metal composition and hand composition using foundry type.

Tschichold based his design for Sabon roman on a font engraved by Garamond, and Sabon italic on a font by Granjon. It was first used in 1966 and has proved an enduring modern classic.